Hanford Nuclear Reservation, where the plutonium for the Nagasaki bomb was produced. Pritikin lays out her material methodically, providing the scientific, medical, legal, and historical components important to the readers' full understanding."

Norma Field, professor emerita, East Asian languages and civilizations, University of Chicago, and author of *In the Realm of a Dying Emperor: Japan at Century's End*

"*The Hanford Plaintiffs* is an extraordinary and unique exposé of the human results of deliberate releases of huge quantities of radioactive isotopes from the Hanford reactors and nuclear complex over many years of operation."

Helen Caldicott, MD

"*The Hanford Plaintiffs* is a unique document: it is a joint effort of the plaintiffs themselves; denied their opportunity to tell their stories in a court of law, their suffering, and their lives downwind, had largely become invisible until now. The Hanford workers and their families and neighbors were deemed expendable by the US government in the national quest for nuclear superiority. For the first time, thanks to the work of Trisha T. Pritikin, we can meet the people who lived through this horror and hear the stories of their pain, their bravery, and their dignity."

Robert (Bo) Jacobs, professor at the Hiroshima Peace Institute and the Graduate School of Peace Studies of Hiroshima City University and author of *The Dragon's Tail: Americans Face the Atomic Age*

"Trisha Pritikin's compelling chronicling of the tragic suffering and deaths of the Hanford community and downwinders, the deliberate guinea pigs and sacrificial victims of the US Manhattan Project secret development of the plutonium weapon destined for Nagasaki, and the United States' ongoing race for global dominance in nuclear weaponry, is essential reading for all concerned with humanity and justice."

Jennifer Allen Simons, founder and president of The Simons Foundation, Canada

"Timely and compelling, with the experiences and voices of impacted people at its core, *The Hanford Plaintiffs* is one of the most important works on Hanford to date. These twenty-four oral histories, coupled with Pritikin's eloquent and accessible analysis of nuclear history, dose reconstruction science, and toxic tort law, make this book essential reading for citizens and professionals alike."

Sarah Fox, author of *Downwind: A People's History of the Nuclear West*

"In *The Hanford Plaintiffs*, Trisha Pritikin reveals the breadth and depth of the devastating health effects of the Hanford site on local residents and the extent to which they were deceived and misled by the US government. A downwinder herself, she presents a thorough account of those whose lives were unknowingly impacted by radioactive and toxic contamination. This book is an urgent reminder of why American citizens must remain diligent and hold our government accountable for the truth we deserve regarding nuclear contamination, especially when lives were—and continue to be—at stake."

Kristen Iversen, author of *Full Body Burden: Growing Up in the Nuclear Shadow of Rocky Flats*

"Irradiated lambs, mysteriously paralyzed children, debilitating cancers: this is the all-too-real nightmare of America's 'forgotten guinea pigs,' the inland Northwest downwinders who have been irreparably harmed by atomic fallout. Thoroughly researched and expertly organized by Trisha T. Pritikin, this is a chilling, startling, and crucial book; no doubt its unflinching truth will inspire change."

Sharma Shields, author of *The Cassandra: A Novel*

"Trisha Pritikin introduces us by name and storytelling to the nuclear justice advocates who demand US government acknowledgement for the ways that radiation exposure upends every aspect of their lives. I will use this book with my students so they can consider the full range of health and environmental dangers connected to our nuclear industries."

Holly M. Barker, principal lecturer, Anthropology Department, University of Washington, and commissioner for the Republic of the Marshall Islands' National Nuclear Commission. She is author of *Bravo for the Marshallese*, and coauthor of *Consequential Damages of Nuclear War*.

THE HANFORD PLAINTIFFS

THE
HANFORD
PLAINTIFFS

VOICES FROM THE FIGHT FOR
ATOMIC JUSTICE

TRISHA T. PRITIKIN

With a Foreword by Richard C. Eymann and Tom H. Foulds
and an Introduction by Karen Dorn Steele

 UNIVERSITY PRESS OF KANSAS

Published by the University Press of Kansas (Lawrence, Kansas 66045), which was organized by the Kansas Board of Regents and is operated and funded by Emporia State University, Fort Hays State University, Kansas State University, Pittsburg State University, the University of Kansas, and Wichita State University.

Library of Congress Library of Congress Cataloging-in-Publication Data

Names: Pritikin, Trisha T., author.
Title: The Hanford plaintiffs : voices from the fight for atomic justice / Trisha T. Pritikin ; with a foreword by Richard C. Eymann and Tom H. Foulds and an introduction by Karen Dorn Steele.
Description: Lawrence : University Press of Kansas, 2020. | Includes index.
Identifiers: LCCN 2019025718 | ISBN 9780700629039 (cloth) | ISBN 9780700629046 (paperback) | ISBN 9780700629053 (epub)
Subjects: LCSH: Hanford Works—Trials, litigation, etc. | Hanford Site (Wash.)—Trials, litigation, etc. | United States. Department of Energy—Trials, litigation, etc. | Toxic torts—Washington (State)—Hanford Site. | Radioactive waste sites—Law and legislation—Washington (State)—Hanford Site. | Radiation victims—Legal status, laws, etc.—Washington (State)—Hanford Site.
Classification: LCC KF228.H285 P75 2020 | DDC 343.7307/862345119—dc23
LC record available at https://lccn.loc.gov/2019025718.

British Library Cataloguing-in-Publication Data is available.

Printed in the United States of America

10 9 8 7 6 5 4 3 2 1

The paper used in this publication is recycled and contains 30 percent postconsumer waste. It is acid free and meets the minimum requirements of the American National Standard for Permanence of Paper for Printed Library Materials Z39.48–1992.

To the civilians caught in the fallout

CONTENTS

Dozens of books and hundreds of articles have been written about the nuclear proliferation brought about by the Manhattan Project and contemporaneous projects in the Soviet Union in the 1940s. As lawyers who spent more than two decades representing victims of radioactive contamination downwind from Hanford, we had the opportunity to meet and speak with hundreds of them. Our fight was against the US Department of Energy, but for the victims the fight was always trying to find a way to deal with injuries that ranged from fear of cancer to constant medication and abnormal life to the worst outcome—dealing with imminent death as the result of living in the shadow of Hanford.

When asked to coauthor a Foreword to this book, we knew there would be no way for us to adequately describe the magnitude of the suffering that we saw over more than twenty-four years of involvement in the infamous In re Hanford litigation. Ms. Pritikin has obviously devoted hundreds, if not thousands, of hours to documenting victims' stories, describing what it is like to live one's life as a downwinder. To our knowledge, no one has previously focused on that aspect of the Hanford legacy.

We were asked if we could provide the names of some of our dearest clients out of the more than one thousand that we represented in the hope that they would agree to an interview with the author of this book. It was ironic, and perhaps saddening, that when we went to our lists, we found that a number of them had passed away, some due to various cancers and others due to natural causes. As we pulled up the old files contained in hundreds of boxes of documents involved in the litigation, we came upon photographs of clients who had died without ever receiving any compensation from the United States and its contractors for their exposures to radiation when they were just babies, toddlers, and sometimes as old as in their teens. Every one of them had a tragic story about how their life had been affected by the many conditions and injuries that radiation can cause.

Because attorney-client privilege remains after a case is concluded, we wrote to those still living to ask whether they would give permission to be interviewed by Ms. Pritikin, herself a Hanford downwinder. Not a single one declined the invitation, for they knew the real truth had never been

Tom Foulds (*left*) and Richard Eymann (*right*), attorneys for the plaintiffs, *In re Hanford*.

revealed in the litigation—at least the real truth about the pain, anguish, and suffering they had endured.

It is our hope that as you read this book, what these stories reveal will lead to global reduction of the incredible risk and the major dangers, not only of nuclear warfare but also of the existing radioactive contamination at many nuclear bomb–making facilities that still need to be cleaned up in this country. It cannot be denied, even by the "in denial" US Department of Energy, that the 580-square mile Hanford is by far the worst of these facilities. It is now estimated that it will cost taxpayers $242 billion to complete the cleanup (which includes fifty-six million gallons of radioactive waste) with a new target date of 2079. As we write this Foreword, our government and certain contractors are continuing the demolition of the plutonium finishing plant at Hanford, and thousands of current downwinders are living with new concern and actual fear that airborne plutonium is escaping the demolition work area and possibly planting cancer time bombs in their children and even themselves.

Richard C. Eymann, Attorney at Law
Tom H. Foulds, Attorney at Law (ret.)
Counsel for the Plaintiffs, *In re Hanford Nuclear Reservation Litigation*

I embarked upon this project in order to create an enduring record of the human toll of Hanford operations through a compilation of the stories of personal injury plaintiffs in the Hanford downwinders' consolidated action, In re Hanford Nuclear Reservation Litigation (In re Hanford).

In 1990, a government report confirmed that, beginning from the startup of operations in late 1944 and for more than four decades thereafter, the Hanford nuclear facility in southeastern Washington State secretly released radiation throughout vast regions of the US Pacific Northwest. Once the contents of the report became public knowledge, nearly five thousand people who had lived downwind and downriver of Hanford filed personal injury toxic tort claims against the facility's former contractors, asserting that cancers and other serious illnesses that they now suffered were the result of exposure decades earlier, primarily while in utero, during infancy, and in childhood, to the ionizing radiation that Hanford had released into the air and to the waters of the Columbia River.

Remarkably, jury trials did not take place until fifteen years after the majority of downwinders' claims were filed. In 2005, six plaintiffs, randomly chosen to serve as bellwethers (representative plaintiffs), testified on the debilitating consequences of autoimmune thyroid disease and thyroid cancers that had, in some instances, metastasized to other organs. The stories of personal injury in the remaining nearly five thousand Hanford downwind plaintiffs who had filed claims against former Hanford contractors remained out of the courtroom and, consequently, out of the public eye. It was advantageous to the defense that courtroom testimony was restricted to the stories of only a small number of representative plaintiffs, as this restriction kept the true magnitude of the harm caused by the radiation released from Hanford operations from becoming public knowledge.

In late 2015, at around the time I shouldered this project, meager sums had just been delivered to the final group of plaintiffs who were proffered settlement in the litigation. With In re Hanford coming to an end, the public would soon forget the suffering of the downwinders of the Pacific Northwest. It had been nearly thirty years since declassified Hanford records first revealed the facility's secret radioactive legacy. Karen Dorn Steele—the investigative reporter who broke the Hanford story, wrote the first accounts of cancer and other potentially radiogenic disease in the downwinders, and covered the legal maneuverings of In re Hanford over many years—had retired from Spokane's Spokesman-Review six years earlier. Only a few local papers mentioned the final settlement of the twenty-four-year litigation in passing. The downwinders continued to suffer and to die, but their stories no longer made the news.

I knew that unless the plaintiffs' stories were made known, the public would remain unaware of the harm that Hanford operations had likely caused to civilian populations downwind. To make these stories known, I first had to locate former downwind plaintiffs willing to be interviewed. I decided to contact two of the downwinders' attorneys, Tom H. Foulds of the Hanford Litigation Office in Seattle and Richard C. Eymann of Eymann Allison Hunter Jones in Spokane. Between them, Foulds and Eymann had represented over a thousand Hanford plaintiffs. I described my vision for the project to both attorneys, and they agreed to distribute a letter to former downwinder clients who they thought might be interested in becoming involved. Nearly everyone contacted decided to participate. If not for the support of attorneys Foulds and Eymann, I would not have had the opportunity to capture the stories that are presented within these pages. I am very appreciative of this support.

Once the project was a bit further along, it found a home with the University Press of Kansas. An external academic reviewer for the press suggested that I interweave into the manuscript the story of people exposed to fallout from aboveground atomic tests detonated from 1951 through the early 1960s at the Nevada Test Site (NTS). The reviewer felt that the reports of leukemia and other cancers in communities downwind of the iconic mushroom clouds produced by these tests would be far more familiar to the public than the parallel accounts of cancers and other disease in communities downwind of Hanford.

Initially, I resisted this suggestion, continuing to focus on health damage in communities downwind and downriver of the Hanford facility. I nevertheless began to look for commonalities in the stories of the two downwind populations. I learned that in the years following the start of aboveground testing, alarming numbers of children in small towns north and east of the test site had been diagnosed with leukemia. These reports mirrored the accounts of children in the Hanford region stricken with leukemia in the early years of Hanford operations. Epidemiological studies had found a prevalence of thyroid cancer in children downwind of the NTS following exposure to the radioactive iodine in atomic test fallout. Thyroid cancer had also been found in people exposed as children to airborne radioactive iodine released from Hanford's chemical separations plants. Other cancers and radiogenic diseases in the two downwind groups were identical.

Declassified documents revealed that the Atomic Energy Commission (AEC) and its successor, the Department of Energy (DOE), had concealed decades of off-site radiation releases from the Hanford facility. The government had failed to monitor exposures to Hanford's downwind and downriver communities and to communities downwind of the NTS. Instead, while radioactive fallout blanketed their communities, both populations were assured repeatedly by government officials that they were in no danger.

I began to understand that the abundance of suffering within the two exposed populations did not constitute two separate stories. Instead, the downwinders were all part of an atomic-age humanitarian catastrophe caused by the actions of the US government, leaving in its wake a vast number of civilian victims deserving of compensation and care.

The two downwinder populations turned to the courts for relief, filing personal injury claims in mass toxic tort litigation. The NTS downwinder litigation, Irene Allen v. US, was filed in 1984, while In re Hanford Nuclear Reservation Litigation claims were consolidated in 1991.

Mass toxic tort litigation can be a highly impersonal experience for plaintiffs. These cases often involve thousands of plaintiffs—far too many to permit face-to-face attorney-client relationships. Many plaintiffs never even meet their attorney. Most plaintiffs never have their day in court, deprived of the closure that publicly testifying on the personal impact of exposure-caused disease and disability can bring. Only the few plaintiffs randomly chosen to serve as bellwethers ever go to trial, before a jury in some cases and before a judge sitting without a jury in others.

I would have liked to capture more of the In re Hanford plaintiffs' stories. Not only would I have been honored to meet more of my fellow downwinders, but with each new story that is added to the last, the true extent of harm suffered downwind of the Hanford facility is brought further into the light.

May our losses serve as a catalyst for change, and may our stories help move this world toward a nuclear weapons–free future.

—Trisha T. Pritikin

Net proceeds from this project will be donated to the work of Consequences of Radiation Exposure (CORE), a Washington State and IRS 501(c)(3) nonprofit organization. The primary mission of CORE is to advance and disseminate understanding of the human toll of exposure to ionizing radiation from uranium mining, milling, or transport; nuclear weapons production, testing, or use in warfare; nuclear reactor off-site releases; and related radiation exposures. CORE hopes to found an international museum focusing on issues relating to the health effects of ionizing radiation. For more on CORE, visit http://www.corehanford.org.

ACRONYMS

ACHRE	President's Advisory Committee on Human Radiation Experiments
AEA	Atomic Energy Act
AEC	US Atomic Energy Commission
ATSDR	US Agency for Toxic Substances and Disease Registry
BCC	basal cell carcinoma
BEIR	Biological Effects of Ionizing Radiation
C.O.R.E.	Consequences of Radiation Exposure
CDC	US Centers for Disease Control and Prevention
CNS	central nervous system
COPD	chronic obstructive pulmonary disease
CT scan	computerized tomography scan
DDT	dichloro-diphenyl-trichloroethane
DOD	US Department of Defense
DOE	US Department of Energy
EEOICPA	Energy Employees Occupational Illness Compensation Program Act
EPA	US Environmental Protection Agency
FFTF	Fast Flux Test Facility
FOIA	Freedom of Information Act
FTCA	Federal Tort Claims Act
GE	General Electric
GERD	gastroesophageal reflux disease
HEW	Hanford Engineer Works
HW	Hanford Works
HEAL	Hanford Education Action League
HEDR	Hanford Environmental Dose Reconstruction Project
HHIN	Hanford Health Information Network
HIDA	Hanford Individual Dose Assessment Project
HMMP	Hanford Medical Monitoring Program
HTDS	Hanford Thyroid Disease Study
I-131	iodine I-131, a radioisotope of iodine
ICRP	International Commission for Radiological Protection

IOM	US Institute of Medicine
IWMF	International Waldenstrom's Macroglobulinemia Foundation
JIA	juvenile ideopathic arthritis
µCi/d	microcuries/day
MRI	magnetic resonance imaging
MS	multiple sclerosis
NAS	US National Academy of Sciences
NCI	US National Cancer Institute
NCRP	US National Committee for Radiological Protection
NPG	Nevada Proving Grounds
NTS	Nevada Test Site
NWRHA	Northwest Radiation Health Alliance
ORERP	Off-Site Radiation Exposure Review Project
P-AA	Price Anderson Act
PPG	Pacific Proving Grounds
PTH	parathyroid hormone
PUREX	Hanford plutonium uranium redox extraction
RECA	Radiation Exposure Compensation Act
REDOX	Hanford Reduction-Oxidation plant
RESEP	Radiation Exposure Screening and Education Program
SEC	Special Exposure Cohort
TSH	thyroid-stimulating hormone
USPHS	US Public Health Service
WBC	Whole Body Counter
WM	Waldenstrom's macroglobulinemia
WPPSS	Washington Public Power Supply System

INTRODUCTION
Karen Dorn Steele

Hanford's closely guarded secret, the manufacture of plu-
tonium near a remote bend in the Columbia River during
World War II, was first revealed in a banner headline after
Nagasaki was leveled: "It's Atomic Bombs."

That triumphalist 1945 story in the Richland, Washington, newspaper extolled Hanford's contribution to the war
effort and its role in forcing a Japanese surrender after the
Hiroshima and Nagasaki bombings. For decades, no subsequent headline told the rest of the story: that the nuclear
arms race, conducted in secrecy, had also claimed victims on
American soil, including from Hanford.

Those victims include Utah sheep who suffered gruesome deaths after atomic bomb tests in the 1950s; people
who developed cancer after atmospheric nuclear tests in the
Marshall Islands and Nevada; and civilians living near Hanford and other nuclear weapons sites who suffered cancers
and other illnesses decades after exposure.

Trisha Pritikin's important new book gives voice to the
little-known victims who sued Hanford's plutonium-manu-
facturing contractors in 1990 after the truth about Hanford's
dangerous emissions had been forced by newspapers and
activist groups from a reluctant federal government.

There are two narratives to this story: unveiling Hanford's
secret past and then following a federal trial over a quarter
century that left many injured people embittered, some with

Karen Dorn Steele.

small settlements but with no government apologies. I played a lead role in the first Hanford narrative after I was recruited away from an on-air news show on public television to join the Spokane local newspaper in 1982 as an environmental and special projects reporter.

The Reagan administration had recently restarted an old Hanford plutonium plant in a renewed nuclear arms buildup. At the same time, the federal government was searching for a safe place to bury a huge cache of highly radioactive spent fuel rods from US commercial nuclear power plants.

My newspaper, the *Spokesman-Review* of Spokane, Washington, is a two-hour drive northeast of the Hanford complex. I was well trained in the use of historical documents, with history degrees from Stanford and Berkeley, and had been tutored in national politics during a stint in Washington, DC, as a congressional intern. I'd reported on the Pacific Northwest for a decade on KSPS-TV in Spokane before joining the newspaper.

Seeking a way to evaluate whether Hanford's underlying geology was safe for the fuel rod burial project and curious about past environmental accidents, I turned to the Freedom of Information Act (FOIA) to request reports on Hanford accidents and possible off-site releases.

The government's response to my requests was stunning. While a series of recent environmental reports had routinely been made public, Hanford officials told me that many documents from the 1940s, 1950s, and 1960s—the years of maximum plutonium production—were still classified under national security restrictions. Without these reports, nobody could determine whether radiation had escaped from Hanford through water and air or whether people had been harmed by routine releases or accidents.

I learned that two federal agencies—the US Environmental Protection Agency (EPA), created by President Richard Nixon in the 1970s to protect the nation's air and water, and the US Department of Energy (DOE), whose nuclear weapons makers had worked in classified programs since the 1940s—were at loggerheads over Hanford's history.

The EPA, armed with legal authority under Superfund laws, was under pressure to study Hanford pollution to protect the Columbia River. The DOE was opposed to disclosure in order to protect the Reagan administration's aggressive nuclear weapons buildup, including the restart of the plutonium uranium extraction plant (PUREX) plant at Hanford.

Journalists, environmental groups, and other activists turned to the federal courts to obtain the early Hanford documents. Attorneys for my newspaper argued that the Freedom of Information Act, a sunshine law passed after the abuses of Watergate in the Nixon administration, allowed details of Hanford operations to be declassified and released.

While this information battle was playing out, an outspoken third-generation farmer living close to the Hanford plutonium plants invited me in 1984 to visit Mesa and Ringold, small towns across the Columbia River from Hanford. Tom Bailie of Mesa told me he'd long suspected that there had been accidents and radiation releases from the nuclear reservation. So had his neighbors. While water, milk, and soil in the area had been monitored by Hanford officials, there were no health studies of the people Bailie called "Hanford downwinders."

Hanford officials told me that they wouldn't expect to find anything because the plant operations had been safe. My editors—at first—were skeptical because of those official assurances.

Bailie's white farmhouse was within what locals called "the death mile," where only one of ten farms had escaped cancer. One couple, Juanita and Leon Andrewjeski, showed me a "death map" they'd compiled to track the fate of their neighbors. It was marked with Os for cancer deaths and Xs for early heart attacks. It was hardly sophisticated science, but it showed their concern over the myriad early heart attacks, cancers, and other illnesses they'd observed. The Andrewjeskis lived in the ten-mile emergency evacuation zone around Hanford's lone commercial nuclear reactor. They'd recently been given a radio to alert them in case a nuclear accident required an evacuation, but they'd lived on their farm since 1954 with no earlier-warning system in place.

Leon Andrewjeski, now dead, recalled an incident in the early 1960s in which government trucks from Hanford, with no notice, had suddenly lined the road outside his farm. Men with Geiger counters fanned out into his fields, inspecting the soil. The men didn't tell him what they were looking for, and the family never got an explanation.

Another Mesa farmer, Bob Love, had worked at Hanford as a technician from the 1940s through 1969. He recalled accidents in the 1960s that were withheld from the public, including radiation releases from the stacks of several facilities: Z-Plant, a plutonium production facility; PUREX, a plutonium reprocessing plant; and a reduction-oxidation (REDOX) plant. Love recalled an incident where flakes of radioactive ruthenium went up the stack of the REDOX plant in the late 1950s. He said radiation teams picked up the flakes to the banks of the Columbia but didn't cross the river to inspect his neighborhood.

Don Worsham, another Mesa old-timer, lived in the "death mile" populated by many of the homesteaders who had settled there in 1954 when the Columbia Basin Irrigation Project opened up more land to farming. He pointed out his kitchen window toward Hanford.

"I've always contended that this thing over there is causing it. Most of the people in this area aren't living to a very old age," Worsham said.

Nels Allison recounted a gruesome story of deformed lambs born in his lambing shed in the winter of 1961. They were mummified, their bones rigidly set, many without eyes, mouths, or sex organs. The birthing ewes died as well, their hair falling out in clumps.

"They looked like little demons," he said of the dead lambs. More than one hundred other lambs on at least six nearby farms were also born severely defective that year. Allison said his neighbor Lyle Taylor, who died in a farm accident before my visit, had lost the most, including thirty-three born in one day. The farmers didn't report the strange livestock deaths.

Returning to Spokane, I requested documents that might corroborate these stories. The monitoring reports that I eventually received showed reason for the farmers' concerns. The Atomic Energy Commission (AEC), predecessor to the DOE, was aware of the possible contamination of farms at Ringold and at Riverview, a farming area farther south along the Columbia. General Electric, which ran Hanford for the AEC, noted, "Some of the [plutonium] reactor effluent radionuclides can be traced through the irrigation water to milk and other farm products." Nuclear secrecy kept these problems from the farmers. Those in the know could not speak out.

F. R. Chen, a farmer with a doctorate in chemistry, reared his family on the Ringold bluff. In 1962, when he worked for the Pasco water district, Chen read a confidential AEC memo warning of high levels of radiation in Columbia River whitefish that had eaten radioactive moss downstream from the N reactor.

"After I read those confidential memos, I said no way am I going to eat fish anymore," he told me. But Chen couldn't warn his neighbors because he would have lost his security clearance.

My initial story on the farm families was published on July 28, 1985, on page one of the *Spokesman-Review*. It featured a photo of Tom Bailie standing chest-deep in corn, with Hanford in the background. Its title was "Downwinders—Living with Fear." A companion story on the mysterious sheep deaths, which had also occurred in Utah during open-air nuclear bomb tests, was titled "The Night the 'Little Demons' Were Born."

Immediately after my story was published, Hanford officials moved to downplay the downwinders' concerns.

"The Energy Department hasn't done off-site studies because we wouldn't expect to see anything," said Don Elle, at the time chief of Hanford's radiological and environmental safety branch. That statement was highly misleading. Later, I would learn of the AEC's deceptive efforts to discredit Utah sheepherders' accounts of similar sheep deaths during radioactive fallout from the most hazardous atomic bomb tests—the Upshot-Knothole series in 1953—and the role that Hanford scientists who secretly conducted studies of iodine-131's impact on sheep played in that deception.

In her book, Pritikin draws parallels between the Utah and Hanford sheep deaths and outlines decades-long government efforts to deny exposed civilians downwind of the Nevada Test Site (NTS) and Hanford any legal remedies for their illnesses and suffering.

My mid-1980s stories on the health concerns of Hanford's neighbors, plus other reporting on safety hazards at Hanford's recently restarted PUREX plant (including in the *Seattle Post-Intelligencer* and the *Seattle Times*), put additional pressure on the government to release Hanford's historic documents.

In 1985, two activist groups, the Hanford Education Action League (HEAL) of Spokane and the Environmental Policy Institute of Washington, DC, filed a broad request under the Freedom of Information Act for Hanford's emission records for the 1940s and 1950s. My newspaper separately requested the reports.

Six months later, the DOE finally conceded to public demands, announcing that it would release nineteen thousand pages of historic reports. On February 27, 1986, Hanford operations manager Michael Lawrence stood next to a tall stack of declassified papers at a crowded press conference in Richland. Declining to answer whether Hanford had caused harm in its early years of operations, Lawrence said he would "leave it to others" to evaluate Hanford's history.

The documents revealed a dramatic story: The government had lied about Hanford's safety. Its Cold War–era operations had emitted more radiation than any other US nuclear production facility—potentially harming tens of thousands of people.

Arguably, the most disturbing document disclosed that day in Richland was an account of a secret military experiment, the "Green Run," that deliberately exposed people from Walla Walla to Spokane to radioactive iodine-131. This by-product of plutonium production is absorbed by the thyroid gland and can cause cancer or other malfunctions decades later. I was the first to report the Green Run story on March 6, 1986, a week after Lawrence's press conference. It was described in a March 1950 quarterly monitoring report that had been kept classified for thirty-six years.

In December 1949, Hanford purposely ran its plutonium plants without health-protecting filters, ejecting "green," or short-cooled, nuclear gases in an effort to emulate what its operators thought the Russians under Joseph Stalin might do to rush the nuclear bomb program they'd accelerated that year. The experiment, conducted by General Electric's Nucleonics Department for the AEC, reprocessed 1 ton of uranium over a 24-hour period that had been cooled for only 16 days. Uranium was normally cooled for 83 to 101 days before reprocessing.

The Green Run released into the atmosphere 5,500 curies of iodine-131 and other fission products, which spread in a two-hundred-by-forty-mile plume from The Dalles, Oregon, to Spokane. There was no public health warning. (In contrast, when the 1979 Three Mile Island accident released fifteen to twenty-four curies of radioactive iodine into the Pennsylvania countryside, people around Harrisburg were evacuated, and milk was impounded.) The resulting contamination from the Green Run exceeded the AEC's laxer 1940s-era "tolerance thresholds" for iodine-131 on vegetation by eleven thousand times on Hanford property and by hundreds of times in some Hanford-area communities.

Allen Conklin, a former Hanford official who had moved to Washington State's office of radiation protection, told me that the early AEC "safety" standards "often exceeded our emergency action levels today."

I turned once again to the Freedom of Information Act for more Green Run documents, eventually learning that it had been a joint AEC-US Air Force operation and that military airplanes had secretly monitored the huge radiation plume that spread over the inland Northwest on the day of the 1949 experiment.

Under public pressure, the DOE released additional information. In September 1987 we learned from ten thousand more pages that radioactive iodine-129 had been detected in some of the deep confined aquifers under Hanford and in farmers' wells across the Columbia River.

The pattern was becoming clear: Hanford officials had chosen to keep secret the routine releases, embarrassing accidents, and nuclear experiments with dangerous public health risks. It was the same pattern of government deception in the 1950s that was laid bare in a 1979 congressional hearing in Salt Lake City, where the AEC's misleading denials of the health dangers of atomic testing were exposed. As former US interior secretary Stewart Udall asserts in his 1994 book, The Myths of August, the AEC's cover-up of the health impacts from the disastrous Upshot-Knothole tests amounted to "the most long-lived program of public deception in U.S. history."

The Salt Lake City hearing was prompted by an explosive February 1979 study by a Harvard-trained epidemiologist, Dr. Joseph L. Lyon of the University of Utah, who tracked childhood leukemia deaths in southern Utah after the AEC's atmospheric bomb tests. Lyon's study, later published in the New England Journal of Medicine, revealed that leukemia deaths among children downwind of the tests were two and a half times higher than the normal national death rate for childhood leukemia. Seeking a similar epidemiological review of the Hanford emissions and their potential health effects, an independent panel of scientists, including Washington and Idaho officials and representatives of three Indian tribes near Hanford, was convened in 1987.

On July 11, 1990, six years after Tom Bailie's neighbors had first spoken to me about their worries, Energy Secretary James Watkins held a high-profile press conference in Washington, DC. Watkins, a former admiral, admitted for the first time that Hanford releases in the late 1940s had been large enough to create a health risk to people living near Hanford.

"The implications . . . are serious. We do have to go back and look at what happened to people," he said. "We don't know now who was in the right spot at the wrong time."

This was an about-face for Watkins, who had earlier advised the Reagan administration against disclosure, saying public knowledge of the potential health damages of nuclear weapons production would threaten the government's renewed nuclear weapons buildup.

After Watkins's admission, Bailie and other downwinders held their own press conference in Richland, describing their feelings about a government that put their health at risk while making nuclear bombs.

"We are the children who lived out there. We are veterans of the Cold War, but we are getting the truth now. A door has opened," Bailie said.

A few weeks after Watkins's admission, the first of the Hanford down-winder lawsuits was filed. Bailie and Pritikin both became plaintiffs. The legal hurdles for the Hanford downwinders were formidable, as Pritikin's analysis shows.

In January 1986, a month before the first batch of Hanford contamination documents was released, the US Supreme Court rejected an appeal by the Utah sheepherders whose flocks had been killed during the Upshot-Knothole atomic tests. The herders had been granted a new trial after Utah federal judge Sherman Christensen had ruled that the government had committed a "fraud upon the court" in the first trial by presenting misleading evidence denying a link between fallout and the sheep deaths. Pritikin describes how scientists involved in the iodine-131 sheep experiments at Hanford had participated in that cover-up.

The downwinders couldn't sue the US government directly under the doctrine of "sovereign immunity," a powerful protection for the weapons makers that was used by the government in Utah in a federal trial, *Allen v. US*, seeking compensation for 1,192 civilians who lived downwind of the atomic bomb tests between 1951 and 1963 and later developed cancer.

In May 1984, when I was interviewing Tom Bailie's neighbors, Judge Bruce Jenkins of Salt Lake City released his verdict in the *Allen* case, awarding $2.6 million in damages to nine of twenty-four bellwether plaintiffs chosen to test the "toxic tort" claims of the larger group of *Allen* plaintiffs over the leukemia they had developed after the bomb tests. Jenkins ruled for the plaintiffs who could show that the dangerous atomic fallout was a "substantial factor" in their later cancers. It was a less rigorous test than the one that would be chosen later for the Hanford downwinders, who were required to prove at trial that "but for" their Hanford exposure, they would not have gotten cancer.

Congress's reaction to the Jenkins verdict was swift. Sen. John Warner of Virginia, a defense hawk, attached an amendment to a 1984 defense rider substituting the US government's Federal Tort Claims Act as the only remedy for the *Allen* plaintiffs in the NTS litigation.

In 1987, the Tenth US Circuit Court of Appeals reversed Jenkins's ruling, saying the government's "sovereign immunity" under the Federal Tort Claims Act barred the Utah plaintiffs from any compensation. In January 1988, two years before the government admitted that it had put people at risk from Hanford emissions, the US Supreme Court declined to hear the Utah downwinders' appeal. The government's legal defenses, coupled with Congress's willingness to shield the bomb makers from legal liability, had made it nearly impossible for exposed people to prevail in court.

After the Supreme Court's rejection of the *Allen* appeal, a prominent Seattle law firm, Schroeter, Goldmark & Bender, dropped out of the developing Hanford litigation, saying it was highly unlikely that the plaintiffs could win their case. My newspaper, the *Spokesman-Review*, published a prescient editorial on July 17, 1990, on the likelihood that the government would never fully account for its harm to the Hanford downwinders: "If the government drags its feet long enough, and makes its studies narrow enough, just about all of the victims will be dead before compensation can be arranged."

Barred from directly suing the government, lawyers for the Hanford downwinders sought compensation under the Price-Anderson Act, a 1957 amendment to the Atomic Energy Act that allowed jury trials against private contractors responsible for a "nuclear incident," including the contractors who had run Hanford for the government during World War II and the early years of the Cold War.

The case, *In re Hanford*, lasted twenty-five years—one of the longest civil suits in the history of the federal courts in eastern Washington. Many plaintiffs died as the case lurched forward, and the final settlement reached in 2015 left most survivors dissatisfied and bitter.

US taxpayers paid an undisclosed amount for the downwinder settlement. They also paid over $80 million in legal fees to the law firms defending the nuclear contractors whose emissions had harmed the civilian populations near Hanford. The US Treasury was obligated to pay the contractors' legal fees under a World War II agreement to indemnify them for the risky and uncertain business of making plutonium.

The outcome of the Hanford litigation was influenced by an important political battle: Who could be trusted to study Hanford's impact on its neighbors? Downwinders and other groups critical of Hanford's legacy of secrecy and denial and mindful of what had happened to the Nevada downwinders wanted the health and radiation dose studies to be conducted independently, not under the direction of the government's nuclear weapons program. That didn't happen.

In March 1986, the Hanford Historical Documents Review Committee was created by the governors of Washington and Oregon, with additional members from area tribes. It hoped to lead an independent dose reconstruction effort. However, the DOE announced in late 1986 that it would fund its own dose study. Emerging litigation from exposed people was the major reason cited, and the government treated the dose study as "litigation defense," according to documents obtained many years later under discovery in the downwinders' trial.

Battelle Pacific Northwest Laboratory, a nuclear contractor, was chosen to estimate the potential radiation doses to the public from Hanford's operations from 1944 to 1966. The $26 million study was called the Hanford Environmental Dose Reconstruction (HEDR) Project. Battelle's first numbers were released in July 1990 for the years 1944–1947, when a cumulative 450,000 curies of radioactive iodine were ejected into the air from Hanford operations. The researchers said most of the 270,000 people living in the ten counties closest to Hanford in those years in eastern Washington, eastern Oregon, and Idaho likely received low doses (about 1.7 rads), but 13,500 people received a 33-rad dose from drinking milk contaminated with radioactive iodine. That higher dose is 1,300 times above what the DOE currently considers safe for people living near nuclear weapons facilities.

Babies and small children who drank milk from cows eating pasture grass got even higher doses, 650 to 2,900 rads, the report said. It also noted that people living downstream of Hanford received annual doses of up to 1.7 rads of radioactive phosphorus from drinking Columbia River water or eating fish from the river.

The study's final report, released in April 1994, concluded that radiation from Hanford's plutonium operations had spread over a much larger area and exposed thousands more people than had previously been thought. The final study decreased by one-third the estimated doses received by people living near Hanford but increased tenfold the estimates for those living farther away—such as in Spokane, the region's largest city, 130 miles northeast, or the small farm town of Ritzville, 70 miles away. It also estimated radiation doses from five radioactive substances released into the Columbia River between 1956 and 1965, with 1960 being the peak year. The airborne releases of iodine-131 were "significant enough to strongly justify a thyroid dose study," said John Till, a nuclear physicist who headed the Technical Steering Panel for the Battelle study. Congress mandated the thyroid study, which would use the Battelle dose estimates to study affected people.

A confidential critique of the Battelle dose estimates by a prominent scientist was kept from the public for several more years. I first came across it in a report prepared for the downwinders' trial, presided over in its early years by Judge Alan McDonald of Yakima, a wealthy rancher and landowner appointed to the federal court by Ronald Reagan. McDonald had sealed the report, so the public was unaware of it until my story appeared in 1995. (McDonald also declared Battelle's dose reconstruction project off limits to legal discovery in its early years.)

The confidential report was written by Thomas Pigford, a nationally known nuclear engineer from the University of California, Berkeley. It was jointly paid for by attorneys for the Hanford plaintiffs and the defendant contractors. Pigford delivered a bombshell. He concluded that the Battelle dose reconstruction study underestimated radiation doses to people living near Hanford and didn't account for radiation "hot spots" caused by topography.

If Pigford was right that the Battelle doses were inaccurate, it would throw off the results of the Hanford Thyroid Disease Study (HTDS), said Scott Davis, a Fred Hutchinson Cancer Research Center epidemiologist, in my January 1995 story on the Pigford report.

That assessment would prove prescient four years later. After the Hanford revelations, the Clinton administration tried to push forward with medical monitoring for the Hanford downwinders, a first for American civilians exposed to radiation from a US nuclear weapons site.

In March 1997, the Agency for Toxic Substances and Disease Registry (ATSDR) announced that it would provide medical monitoring for up to fourteen thousand people exposed as children to Hanford's radiation releases. The checkups would be for thyroid cancer and other thyroid diseases in people who got an estimated dose of ten rads and above. Ten rads is the dose at which the lifetime risk of thyroid cancer for those exposed as small children is expected to double.

"Even though fifty years have elapsed, these people are still at risk," said Bob Spengler, ATSDR's assistant director for science in Atlanta. The agency is a branch of the US Centers for Disease Control (CDC).

The ATSDR study was opposed by the DOE, which didn't want to pay for it out of Hanford cleanup funds. Under law, DOE was required to fund the study because medical monitoring is a legal requirement of the Superfund statute that governs the Hanford cleanup.

When squabbling over the funding continued, Trisha Pritikin sued in 1998 in federal court in Spokane in an effort to force the DOE to pay for the $12.9 million screening project. Pritikin pointed out that while the DOE claimed that it couldn't afford the study, it had already spent $54.2 million on legal bills for the defendant Hanford contractors to fight the downwinders.

"They are refusing to pay for a mandatory program required under the law," said Seattle attorney Tom Foulds, who represented Pritikin in the ATSDR lawsuit and in the downwinders' litigation.

"This is not about money. It's about justice. The injured people should get an apology, and their health problems should be monitored and treated. Is that too much to ask?" Pritikin said when she filed her lawsuit.

In a show of bipartisanship, Sen. Patty Murray, D-Washington, and Rep. George Nethercutt, R-Spokane, both endorsed the screening program. But the Republican majority on a House-Senate conference committee argued that the program was premature because the National Academy of Sciences had recently recommended against thyroid screening for millions of Americans exposed to bomb-test fallout in the 1950s and early 1960s.

In October 1998, attorneys for the US Department of Justice argued in federal court in Spokane that Pritikin had no right to sue to compel the government to start medical monitoring, citing the government's "sovereign immunity." Foulds disagreed, saying Superfund law required the DOE to pay. He said the 1992 Federal Facilities Compliance Act required the DOE to comply with all federal laws, including Superfund. Before that compliance law was enacted, the government's weapons agencies had often declared themselves exempt from federal pollution laws for national security reasons.

Pritikin lost her legal battle in April 1999 when US District Court Judge Edward Shea of Spokane ruled that the fourteen thousand Hanford downwinders lacked standing to sue the DOE.

Another project, the HTDS, proved detrimental to the downwinders.

The study began after Congress, facing public outrage over Hanford's emissions, authorized it in 1988 and ordered it to be supervised and funded by the CDC. The epidemiological study, conducted by scientists at Fred Hutchinson in Seattle, began in 1991 and investigated whether people living near Hanford had a higher incidence of thyroid disease, the primary health disorder for people exposed to radioactive iodine-131. It relied on the dose estimates in the Battelle dose study. Till, the dose study's director, said the Battelle doses would give the Fred Hutchinson researchers "vital data in determining if a causal link exists between exposures to radioiodine and thyroid disease in the vicinity of the Hanford site."

The National Academy of Sciences also weighed in, approving the continuation of the study after reviewing its pilot phase. The prestigious academy said Hanford was the best site in the nation to study the results of iodine-131 exposure in children. But when Fred Hutchinson reported the draft results of its $18 million study in January 1999 at a press conference in Richland, no elevated thyroid cancer rate was reported. The researchers said they couldn't correlate the Hanford doses with increased thyroid cancer rates.

However, the study of 3,441 people in seven eastern Washington counties did find slightly elevated levels of thyroid abnormalities and an unexplained 20 percent higher-than-normal infant death rate from birth defects and problems late in pregnancy or in the first week after birth. Estimated doses to people in the study were zero to 280 rads. By comparison, in the event of a civilian nuclear plant accident, an evacuation would likely be ordered if a 25-rad dose was expected.

The inconclusive results stunned and angered downwinders, who asked why a dose-cancer correlation wasn't found at Hanford when it had been detected in the Marshall Islands, where residents were exposed to H-bomb fallout; among Utah schoolchildren exposed to fallout in Nevada bomb tests; and in the Ukraine following the 1986 explosion of the Chernobyl nuclear reactor. Project endocrinologist Thomas Hamilton said the exposure conditions at Hanford were "quite different" compared to the other radiation exposure incidents.

At the public meeting where the results of the draft report were announced, Hanford downwinders expressed their fury. Sally Sanders of Kennewick, who had lost her mother and fifty-two-year old sister to thyroid cancer and whose brother also had the disease, faced the researchers with an American flag and a large sign that read, "I don't believe it."

Fred Hutchinson researcher Hamilton, epidemiologist Davis, and statistician Kenneth Kopecky said they were initially surprised that their study had come up with negative results but were confident in its accuracy.

"This study was powerful in the number of people examined and in its design," Davis said. Davis cautioned that the study could not answer whether an individual's case of thyroid disease was a result of Hanford radiation. "People want to know what about me? We can't answer that," he said. However, Davis also asserted that downwinders should be "reassured" by the study results—a comment that drew backlash from many exposed people.

In the week after the press conference, CDC epidemiologist Paul Garbe sought to walk back Davis's sweeping assertion, saying the study results "do not prove there is no link between thyroid disease and iodine-131."

In June, the National Academy of Sciences held a fact-finding hearing in Spokane as part of its peer review of the Fred Hutchinson study and heard another barrage of critical statements from the public.

"The Hanford doses are general estimates at best, and the study is inconclusive. Significant thyroid disease was found, but that message was buried," Pritikin told the scientific panel.

Tim Connor, a Spokane journalist and editorial director of the Northwest Environmental Education Association, said the rollout of the thyroid disease study was a "debacle" that would allow Hanford contractors to "exult in jubilation" while portraying downwinders as alarmists.

After completing its peer review of the draft thyroid study, the National Academy of Sciences had strong criticisms of its own. It flagged a number of errors, including mistakes in some of the dose calculations. Its report was released in December 1999, when a subcommittee of the academy's radiation effects research panel returned to Spokane. While the thyroid study design was sound, it was statistically weak and exaggerated its conclusions, their review concluded. The Fred Hutchinson researchers hadn't disclosed that their uncertainty analysis—a test of the study's power to detect a radiation effect—had failed, the academy critique said. Researchers also had not addressed problems with the Battelle dose estimates that could have skewed the outcome, they noted.

"Shortcomings in the analytical and statistical methods used by the study's investigators overestimated the ability to detect radiation effects," the reviewers said. "Despite these problems, the results of the [study] were presented with unqualified certainty."

The Battelle computer model upon which the Fred Hutchinson researchers had relied had many uncertainties, including how much iodine-131 was released at a given time, which way the wind was blowing, how much of the iodine-131 stuck to vegetation that may have been consumed by cows and transferred to their milk, and how much contaminated milk was consumed by a particular child, said panel chair Roy E. Shore, professor of environmental medicine at New York University School of Medicine.

Several panel members said the Battelle model may have underestimated the iodine-131 releases by a factor of 30 percent, skewing the dose estimates downward. Panel member Sharon Friedman, director of Lehigh University's science and environmental writing program and a professor of journalism, directed interviews with dozens of people about the study's release. She concluded that the Fred Hutchinson scientists made "ill-advised" statements about their study's conclusions before any peer review.

The CDC, funder and supervisor of the HTDS, was also criticized for its decision to publicly release the thyroid disease study as a draft when it hadn't undergone peer review. That decision had been made at CDC headquarters in Atlanta. Garbe, the CDC epidemiology director, said the CDC had decided to release the draft after the National Academy of Sciences had said that it would have to be made public before the academy would review it. The CDC owed a better response to sick people who had been lied to for decades about Hanford's secret radiation releases, journalist Connor said.

"An epidemiological study is a very rigorous test. You are often left with inconclusive results and a community feeling deprived of their experience. . . . [The CDC] did some good work, but they should have been more cautious and thoughtful," he said.

The effect of the bungled rollout of the thyroid study hurt the prospects for medical monitoring for downwinders, said Steve Corker, a public relations adviser to the Hanford Health Information Network (HHIN), a Washington State program to disseminate information to downwinders.

After the release of the inconclusive Hanford thyroid study, ATSDR's initial plans for an ambitious program of medical monitoring were scaled back to an "information and education" program. In early 2001, the CDC prematurely disbanded its advisory panel for the Hanford thyroid study—prompting criticism that the agency would have no continuing oversight to review changes in the study.

The disappointment was not new for the downwinders. In mid-2000, the DOE announced that it was shutting down the Hanford Health Information Network due to budget cuts. Congress had authorized the program in 1991, and $12.7 million had been spent to provide information on Hanford's radiation releases to the public.

That same year, DOE announced that it would establish a program to compensate nuclear workers at Hanford and other US weapons sites up to $150,000 for illnesses caused by their exposure to radiation and chemicals. No similar program was created for civilian victims near the weapons sites.

Pritikin decried the DOE decision to divide workers from their families in government compensation schemes. Her father, Perry Thompson, a nuclear engineer at Hanford, died of metastatic thyroid cancer, and her mother, Lesley Thompson, died of malignant melanoma. Pritikin's infant brother died shortly after his birth in 1947—part of the cohort of mysterious newborn deaths near Hanford in the 1940s and 1950s flagged by the HTDS. If the downwinders exposed secretly as children had been nuclear workers, some would have qualified for compensation under the radiation exposure limits set by the DOE worker program, Pritikin said.

"It's the ethical issue that bothers us downwinders the most," she told me for a March 23, 2000, story.

Earlier efforts to assist the atmospheric test victims were thwarted for years. In 1979, Sen. Ted Kennedy tried to get a law passed compensating Utah sheepherders and the NTS downwinders, but the law was blocked by the Carter administration. In 1980, Sen. Orrin Hatch of Utah unsuccessfully tried to get Congress to approve monetary damages of $50,000 to $150,000 for the Utah downwinders and sheep men. In 1985, Congress acknowledged that atomic bomb testing had caused deaths and illness to exposed people but created a trust fund only for the Marshall Islands, not the NTS, despite Hatch's efforts.

The Hanford downwinders pinned their hopes on a favorable outcome in federal court. But during their twenty-five-year battle, most of their claims were either dismissed or compensated with small sums. The nuclear contractors named in the case were a who's who of the Fortune 500: General Electric, E. I. du Pont de Nemours, Atlantic Richfield Co., United Nuclear, and Rockwell International.

The first judge assigned to the downwinders' litigation was Alan McDonald, the Republican lawyer and businessman from Yakima. As the case got under way, lawyers for the Hanford contractors praised McDonald's decisions, while downwinders observed that the judge appeared hostile to their claims. McDonald "represented the government against us," said Wanda Berg of Republic, Washington. She and her three children were all plaintiffs.

In 1998, McDonald dismissed 4,500 plaintiffs, leaving 20 to 200 people to pursue their claims. He said most of the downwinders couldn't prove that Hanford emissions had doubled their risk of getting cancer, the scientific threshold he chose for the case. Lawyers for the downwinders appealed, saying the standard of proof should be that the Hanford emissions "more likely than not" caused the downwinders' illnesses and that McDonald's decision was premature in the trial's first phase.

The Ninth US Court of Appeals overruled McDonald in June 2002. Lawyers for the defendant contractors appealed for rehearing, and the Ninth Circuit rejected the defense petition in October 2002.

While the Hanford case was on appeal, McDonald made a land purchase that would lead to his recusal. Foulds, Pritikin's lawyer, began examining McDonald's holdings after a March 2000 story I wrote based on public disclosure forms required of federal judges that showed that McDonald owned as much as $39 million in farmland and other assets. Those assets made him the wealthiest federal judge in the Eastern District of Washington. Foulds, a Seattle attorney for the downwinder plaintiffs, decided to look closely at the judge's land holdings. The orchard property that McDonald had purchased in 1999 was just two miles from Ringold, where Tom Bailie's neighbors had been hit with Hanford's heaviest radiation doses. McDonald had certified to a bank as part of the purchase process that the land was free of radiation contamination.

Foulds's court motion questioned McDonald's impartiality for owning agricultural land "squarely within the zones of the highest air-borne concentrations and deposition of the radionuclides released from Hanford." McDonald's land holdings had not been disclosed when he was assigned to the Hanford litigation. Foulds argued in his December 2002 motion that the judge had a financial stake in the Hanford case outcome because the value of his land and crops could be diminished by a verdict for the downwinders.

Three months later, in March 2003, McDonald stepped aside, saying he had done nothing wrong but acknowledging that the orchard issue could be used against him on appeal. The case was reassigned to William Fremming "Frem" Nielsen of Spokane, who was nominated for the bench by George H. W. Bush. Nielsen told the *Spokesman-Review* that he would sell his General Electric stock to avoid any conflicts of interest because General Electric was a defendant in the Hanford case. In a May 2003 order, Nielsen promised an "expeditious, economical and just" resolution to the case, which included nearly five thousand plaintiffs. Nielsen also unsealed the Pigford report, which McDonald had kept under seal for nine years.

The case crept forward for another twelve years. By the time Nielsen took over, $60 million had been paid out to law firms for the defendants, the result of the government's agreement to fully indemnify the private corporations who had signed on as nuclear contractors in the government's weapons program. With no pressure to settle, attorneys for the defendant contractors could prolong the litigation indefinitely—with US taxpayers footing the bills.

In 2004, the defendants once again tried to have the trial thrown out, this time by arguing that the statute of limitations had run out. They said people who had read my July 28, 1985, story, "Downwinders—Living with Fear," on Tom Bailie and his neighbors and then read the media coverage of the DOE's February 27, 1986, release of the declassified Hanford environmental reports should have had "knowledge of causation" by that date, triggering a three-year timeline to file suit.

Lawyers for the downwinders said the proper date triggering the litigation timetable started on July 11, 1990, when the government finally acknowledged that the Hanford emissions were significant enough to cause serious risk of illness. (The first downwinders' lawsuit was filed less than a month later, on August 6, 1990.)

Nielsen sided with the downwinders on the statute of limitations. He also said he wanted to move the case forward by choosing a handful of "bellwether" plaintiffs for jury trials to test radiation exposure claims for downwinders who had developed thyroid cancers and other abnormalities—the same approach used by Jenkins in the *Allen* case in Utah.

In early 2005, before the bellwether cases began, I reviewed copies of documents indicating that Hanford's $27 million dose reconstruction study had been set up at least in part to defend the government against lawsuits by exposed people. The documents, obtained during legal discovery for the Hanford case, contradicted the DOE's assertion that the HEDR Project was simply a work of unbiased science. The documents showed that the US Department of Justice had opposed a dose study as useless "public relations" but changed its mind as soon as the first downwinders' lawsuit was filed in 1990. The Departments of Justice and Energy then treated the HEDR Project, set up in 1988, as a vehicle to provide "litigation defense" to fight claims by exposed people.

The documents showed that some of the Battelle staff in Richland who had worked on the study had also worked for Kirkland & Ellis, the Chicago law firm hired to defend the Hanford contractors. They provided "startling evidence" that the study had been shaped to support the government's litigation defense against the downwinders, including dose estimates that minimized the estimated radiation exposures, Pritikin's attorney Foulds said in a court motion.

Kevin Van Wart of Kirkland & Ellis, representing General Electric, was furious about my story and demanded a change of venue—a request that Nielsen rebuffed. Van Wart denied that the HEDR Project had been set up to favor Hanford contractors.

There were mixed verdicts in the six bellwether trials. Two plaintiffs with thyroid cancer, Gloria Wise and Steve Stanton, born in Pasco and Walla Walla, respectively, in 1944, were awarded a total of $545,000. These verdicts were significant, marking the first time in the nation that a trial jury had found nuclear contractors liable for radiation damage to people exposed to radiation emissions from the government's nuclear weapons program.

The jury rejected the claims of three others with autoimmune thyroid disease: Katherine Goldbloom, whose family farmed near Kennewick; Shirley Carlisle, born in Richland in 1947; and Wanda Buckner, born in Pasco in 1945.

Another bellwether, Shannon Rhodes of Coeur d'Alene, Idaho, was suffering from an aggressive thyroid cancer as she went through two trials. She was raised on a farm near Colfax in the 1940s, when clouds of radiation from Hanford drifted over pastures and tainted milk was consumed by children. In spring 2005, her first trial ended in a 10–2 mistrial. In November 2005, a second jury voted in the defendants' favor, saying Rhodes hadn't proved that Hanford had caused her cancer. Her lawyers had asked for a judgment of $20 million to $30 million for her pain and shortened life.

"I'm heartbroken," Rhodes said after the jury verdict. "These corporations and the government did this to me and half of Washington State, and now they won't be held accountable." Rhodes died in 2011, her life cut short by her disease.

The bellwether plaintiffs were the only downwinders to get an opportunity to tell their stories to a jury. Over the next ten years, some plaintiffs and their lawyers died, others were dismissed from the case, and some agreed to modest settlements.

In 2011, 139 people with hypothyroid disease represented by Oregon attorney Roy Haber accepted settlements of $5,683 each. As of early 2013, 727 claims remained.

Jacklyn O'Neil, an eighty-one-year Otis Orchards woman whose cancerous thyroid gland was removed in 1998, joined the litigation hoping to recoup some of her medical bills. When I wrote about her case in 2013, O'Neil said she was frustrated with her $10,000 settlement offer. "I felt it was a total insult. I think they're just waiting for us to die," she said.

"Congress passed the Price-Anderson Act to assure compensation to victims of nuclear accidents, but the Hanford case shows that many civilian victims of nuclear exposure will die without compensation," said Brian DePew of Los Angeles, who represented hundreds of people with hypothyroidism in the Hanford case.

"It's not encouraging, to say the least," Depew said in 2012 while negotiating settlements for his clients, offered $6,100 each.

In late 2015, a final mediated settlement was reached with the remaining plaintiffs with thyroid cancer and other thyroid abnormalities represented by Tom Foulds. Foulds was the last holdout, fighting to the end for jury trials for his clients.

When the final settlements were announced, defense attorney Kevin

Van Wart insisted that they be kept confidential. There was no further government apology beyond the DOE's acknowledgment in 1990 that Hanford's radiation releases had risked the health of people living downwind from Hanford.

Lawyers for the private nuclear contractors, with a blank check from taxpayers to pay their legal fees, far outspent what lawyers for the downwinders could afford, said Richard Eymann, a Spokane attorney who represented several hundred plaintiffs in the Hanford litigation. Eymann said plaintiffs' attorneys spent less than $10 million on the case, while the nuclear contractors billed the government over $80 million in legal expenses. The government also spent nearly $70 million on dose studies, monitoring, and medical information.

The downwinders weren't fairly compensated by the government, Eymann said. "If we had had the budget they had, it would have been a far different outcome. They used a scorched-earth defense."

For twenty-five years, the Hanford downwinders sought redress for the harm inflicted by a secretive nuclear weapons program. They achieved a legal precedent with a jury verdict for two plaintiffs with high estimated radiation doses and thyroid cancer. But for most, the lawsuit ended with an undisclosed settlement that itself symbolized government secrecy during the nuclear arms race. Trisha Pritikin's book gives some of the plaintiffs in the lengthy Hanford case an opportunity to tell their stories—a chance denied them in federal court.

I

For the survivor who chooses to testify it is clear:
his duty is to bear witness for the dead and for the living.
—Elie Wiesel, Night

An expectant mother drinks multiple glasses of milk every day throughout her pregnancy, faithfully following the latest government-issued nutritional guidelines for a healthy baby. Unbeknownst to her, the milk she drinks is saturated with a radioactive isotope of iodine known as I-131. The radioactive iodine easily crosses the placental barrier, gravely damaging the thyroid gland of her fetus.

The year is 1950.

The fetus is me.

Throughout the US Pacific Northwest, beginning in late December 1944 and for several decades thereafter, thousands of Americans like my mother and me were chronically exposed to destructive levels of ionizing radiation. We have reaped the devastating health consequences throughout our often abbreviated lives.

In this book, I endeavor to give voice to a group of Americans whose stories are largely unknown. These are the stories of innocent civilians poisoned by widespread radioactive fallout, the by-product of plutonium production at the Hanford nuclear weapons complex, a federal facility that operated on the Columbia River in southeastern Washington State without public scrutiny, exempt from most federal, state, and local regulations, for more than forty years.[1]

These are the accounts of only a tiny segment of the thousands of people who lived within Hanford's vast downwind expanse, people placed in harm's way and repeatedly lied to by the US government. These are the stories of one of the first groups of American civilians sacrificed by the United States in its quest for an atomic bomb, a quest that persisted during the Cold War as the relentless pursuit of nuclear supremacy over the Soviet Union. Far more familiar are the kindred accounts of cancer and other radiation-related illness in the Nevada Test Site (NTS) downwinders, civilians who lived within the path of fallout from atomic testing at the continental test site initially known as the Nevada Proving Grounds (NPG) in Nye County, Nevada. Aboveground (atmospheric) atomic tests commenced at the NPG in January 1951, many years after radioactive fallout began to blanket communities downwind and downriver of the Hanford facility.

There are several reasons for the public's far greater familiarity with the stories of the NTS downwinders. Perhaps most importantly, aboveground atomic detonations at the continental test site gave rise to towering mushroom clouds that ascended high into the stratosphere, visible for hundreds of miles. Following the detonations, conspicuous pink- or gray-tinged fallout clouds regularly passed over ranches, herds of grazing livestock, and small towns downwind of the test site.

The manufacturing process at Hanford did not possess the intimations of danger embodied by atmospheric detonations at the continental test site. Hanford's manufacturing process was both invisible and inaccessible to the public, conducted under tight security within the vast federal nuclear site. The airborne and river-borne radioactive by-products of plutonium production released from the Hanford facility were for the most part imperceptible, making it possible for the AEC and its successor, the DOE, to conceal the extraordinary hazards posed to public health by Hanford operations. The chronic off-site release of low-dose ionizing radiation from the facility, often carried out in the dark of night, was not publicly revealed by the DOE until 1986, more than forty years after the start-up of the Hanford facility.

Following the war, the nation's leaders hoped that further testing and development of the country's newly proven atomic arsenal would provide the upper hand in its dealings with Soviet postwar aggression. In 1946, the United States began to test atomic (fission) bombs, similar to the bombs dropped on Hiroshima and Nagasaki, within the Pacific Proving Grounds (PPG) in the mid-Pacific. These tests were followed, beginning in 1952, by far more powerful thermonuclear tests.

After the Soviet Union conducted its first atomic test in August 1949, Americans began to fear what might happen if the Soviets launched a nuclear attack against the United States.[2] By the late 1950s, 60 percent of American children, many of whom regularly took part in "duck-and-cover" drills at school, reported having nightmares about nuclear war.[3]

In January 1951, aboveground (atmospheric) atomic tests began at the NPG. These tests were far less powerful than many of the thermonuclear detonations later conducted within the PPG. That same month, 940 miles north-northwest of the NPG, the Hanford nuclear weapons facility entered its seventh year of operations, its covert release of radioactive and chemical toxins throughout the Pacific Northwest and to the waters of the Columbia River unabated.

All was not well within communities downwind and downriver of Hanford, but at the time, no one had reason to suspect that the secret federal nuclear facility might be to blame. In the mid-1950s, families began to arrive at Lady of Lourdes Hospital in Pasco seeking treatment for critically ill children stricken with leukemia.[4] In 1961, across the Columbia River from Hanford's production reactors and massive chemical separations plants, hundreds of grossly deformed and stunted lambs were born, some stillborn and some surviving only a few days after birth, a heartbreaking disaster that longtime area sheep ranchers had never encountered before. In Eltopia, a small farming town not far from Hanford, Juanita Andrewjeski suspected that something was harming the health of her community. In response to growing reports of cancers in her neighbors and heart attacks suffered by young men working in the fields, she created a map to track the expanding list of victims.

On the rare occasion when the conservative, patriotic community surrounding Hanford questioned whether the behemoth federal nuclear facility might have something to do with all this suffering, Hanford officials would serve up their standard rejoinder: Hanford posed no harm to the public, and the childhood leukemias, cancers, heart attacks, and deaths and deformities in livestock, tragic as they might be, could have nothing to do with Hanford operations.

The devastation that nuclear weapons can cause had been revealed to the American public through the atomic decimation of Hiroshima and Nagasaki in August 1945. Because Americans so dreaded nuclear bombs, they paid close attention to the atmospheric nuclear tests conducted throughout the 1950s in Nevada. The focus of the nation was on the iconic mushroom clouds soaring skyward following detonation, a breathtaking demonstration of the capabilities of the United States in the face of the new dread of nuclear attack by the Soviet Union.

Members of Congress representing regions to the north and east of the test site were keenly aware of the tests, particularly when fallout clouds blanketed their downwind constituents with irradiated debris and microscopic radioactive particles sucked up during the tests. They paid even greater attention when, in the mid-1950s, childhood leukemias and thyroid cancers were first diagnosed within these communities.

A series of congressional hearings convened to examine the health issues reported downwind of the test site allowed grieving downwind communities the opportunity to share, on the record, personal accounts of leukemia and other cancers in families, neighbors, and friends. Much of this heartbreaking testimony was covered by the national media. The litany of suffering brought to light through the downwinders' stories served as the impetus for a congressional report in 1980 identifying NTS downwinders as "forgotten guinea pigs."[5]

I would argue that those of us who were born and raised downwind of the Hanford facility or who frolicked as children in the radioactive waters of the Columbia River during Hanford plutonium production years, and who now bear the burden of radiogenic cancer and other radiogenic disease, are in equal measure America's forgotten guinea pigs.

The stories that follow lay bare the anguish of cancer diagnosis and the debilitating lifelong consequences of the harsh, and sometimes ineffective, surgical procedures, chemotherapy, and radiation treatments that follow such a diagnosis. For the downwinders, life often hinges upon the cataclysmic unknown—will my cancer return; what else will happen to me?

The narratives that follow chronicle lives burdened with cancers; crippling autoimmune disease; neurological disorders; and the deeply personal grief of infertility, miscarriage, stillbirths, and neonatal deaths endured by many of the women who spent their childhood within the radioactive shadow of the Hanford facility.

Tragically, the story does not end with the shattered lives of civilians downwind of Hanford and the NTS. US nuclear weapons facilities, including the Trinity Test Site in New Mexico, the location of the world's first test of an atomic bomb on July 16, 1945, and former nuclear weapons production sites operating during the wartime Manhattan Project and ensuing Cold War era, are known to have exposed downwind communities to a range of airborne radioactive toxins.[6] Compensation to the downwinders, US civilian victims of nuclear weapons, for pain and suffering and other losses relating to cancer and other radiogenic injuries has ranged from meager to nonexistent. The majority of civilians exposed, many during infancy and childhood, have received neither help nor apology from the US government for the extraordinary personal price they have paid for our country's nuclear aspirations.

The stories of America's forgotten nuclear guinea pigs must be heard. Person by person, story by story, these chronicles of lives broken and lives destroyed must be brought into the light. Only then can we begin to understand that as a nation, having turned a blind eye to the human toll of America's nuclear ambitions, we are killing our own.[7]

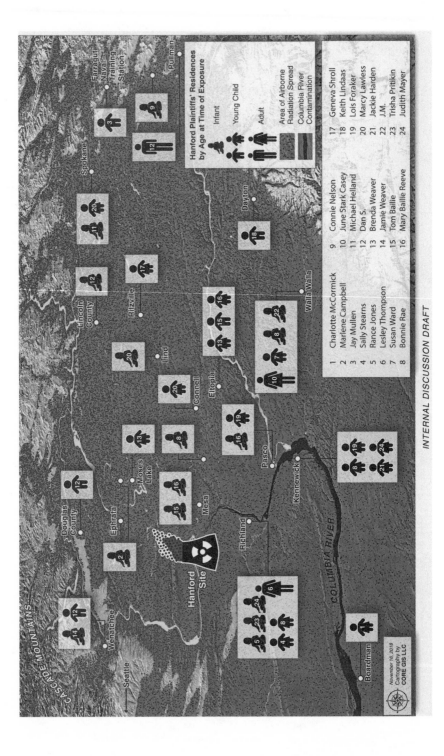

Hanford Plaintiffs' Residences
by Age at Time of Exposure

Infant

Young Child

Adult

Area of Airborne
Radiation Spread

Columbia River
Contamination

1 Charlotte McCormick
2 Marlene Campbell
3 Jay Mullen
4 Sally Stearns
5 Rance Jones
6 Lesley Thompson
7 Susan Ward
8 Bonnie Rae

9   Connie Nelson
10 June Stark Casey
11 Michael Helland
12 Dan S.
13 Brenda Weaver
14 Jamie Weaver
15 Tom Bailie
16 Mary Bailie Reeve

17 Geneva Shroll
18 Keith Lindaas
19 Lois Foraker
20 Marcy Lawless
21 Jackie Harden
22 J.M.
23 Trisha Pritikin
24 Judith Mayer

November 16, 2018
Cartography by
CORE GIS LLC

INTERNAL DISCUSSION DRAFT

2

Hanford's B reactor was the first large-scale plutonium production reactor in the world. It was designed and built by E. I. du Pont de Nemours based on an experimental design by Enrico Fermi.[1] Three plutonium production reactors, known as B, D, and F, were constructed along the Columbia River within Hanford's 100 Area during World War II. By 1955, eight production reactors were in operation along the river. N, a dual-purpose production/power reactor, was built in 1963.

The core of each reactor was composed of blocks of graphite, drilled to allow more than two thousand parallel aluminum tubes to run horizontally from the front to the rear face of the reactor.[2] Thick concrete-and-steel walls surrounding the core provided radiation shielding. The tubes were loaded end to end with fuel slugs measuring 1.6 inches in diameter by 8 inches in length. The slugs, coated with aluminum, were packed with refined uranium-238 (U-238) and a small amount of uranium-235 (U-235), which would easily undergo fission (splitting of the nucleus).

In nature, when the nucleus of an atom of U-235 undergoes fission, most of the neutrons released are captured by atoms of U-238, an isotope of uranium that does not easily fission. The fission process then stops, as there is not enough U-235 to sustain a chain reaction.[3] In Hanford's reactors, U-238 atoms in the uranium fuel (approximately two hundred tons in each reactor) were bombarded by neutrons

Workers laying the graphite core for Hanford's B reactor, Atomic Energy Commission. Photo courtesy of Library of Congress.

from fissioning U-235 nuclei. The graphite core of the reactor served as a moderator, slowing down the speeding neutrons released from the fissioning U-235 nuclei, allowing the nucleus of a U-238 atom to absorb a slow neutron and form uranium-239 (U-239). U-239 rapidly underwent beta decay, transmuting (changing) into neptunium-239, which then underwent a second beta decay, transmuting into plutonium-239. A very small percentage of the original U-238 was transmuted through this process into plutonium-239.[4]

Cooling water from the Columbia River was pumped through the tubes around the uranium slugs.[5] Vertical and horizontal control rods made of boron, silver, indium, and cadmium were able to absorb many neutrons without undergoing fission. The control rods regulated the fission rate of the uranium and prevented Hanford's production reactors from undergoing an uncontrolled, catastrophic chain reaction.

Once maximum plutonium production took place, a remotely operated ram [push rod] drive the irradiated fuel slugs out the back of the reactor into shielded casks filled with water for cooling. Although the irradiated fuel slugs were still solid metal, the nuclear composition of the slugs had changed. Chemical separation would be required to remove the plutonium from the slugs. The length of time that the irradiated slugs were cooled determined how much radioactivity the slugs still held when they were transported to chemical separations and how much gaseous radioiodine and other radioactive by-products would be created during the separation process.[6]

After the cooling period, long tongs remotely transferred the fuel slugs from the water basins into metal transport buckets inside shielded, water-filled casks on railcars built of concrete and steel. Hanford railway workers operating diesel-powered locomotives pushed the cask cars along tracks laid from the reactors to the tunnel entrances of huge rectangular, remotely operated chemical separations facilities ("canyons") in Hanford's 200 Area, a distance of about ten miles.

Within the separation canyons, the irradiated fuel slugs were dumped into nitric acid, dissolving the fuel-cell jacket and releasing the plutonium-239 along with other radioactive by-products. During dissolution, large quantities of I-131 and other radioactive gases and particles were forced, under pressure, up and out of two-hundred-foot-high stacks. Officials were concerned that the topography of the Hanford region could lead to temperature inversions trapping radioactive waste gases released from the plant during routine operations or from a catastrophic reactor explosion.[7] But because construction was already well underway by this time, "the Hanford site had to be accepted with all of its flaws."[8]

Plutonium-239, now converted by the nitric acid to a nitrate compound, went through further chemical processing and was refined into Hanford's final product, plutonium nitrate solution, which was then sent to Los Alamos.[9]

Life in the Sacrifice Zone

The Hanford facility was deliberately sited in the vast desert expanse of eastern Washington, far from any major population centers. Maj. Gen. Leslie R. Groves, the US Army Corps of Engineers officer who oversaw the construction of Manhattan Project sites, including Hanford, was

Atomic Bomb Plant — H.E.W. Process Bldg.
Photo by Robley L. Johnson

Hanford plutonium chemical processing canyon, Atomic Energy Commission.
Photo courtesy of Library of Congress.

extremely concerned that if "because of some unknown and unanticipated factor a reactor were to explode and throw great quantities of highly radioactive materials into the atmosphere when the wind was blowing . . . the loss of life and the damage to health in the area might be catastrophic."[10]

But Hanford needed workers, and those workers had to live somewhere. From the small farming town of Richland, an elite planned community for plant operators, engineers, scientists, and their families was hurriedly constructed. Kennewick and Pasco were the only other towns in the immediate area. In many ways, due to the acknowledged hazards of plutonium production, the "Tri-Cities" of Richland, Kennewick, and Pasco, along with a small number of neighboring farm communities, would become a sacrifice zone should the potential nuclear disaster that so troubled General Groves become a reality, producing an explosion in one or more of Hanford's reactors and showering the people of the area with lethal levels of ionizing radiation.

Hanford's reactors did not have metal shells or safety domes to contain fission products should such an explosion occur.[11] A reactor meltdown, caused by water loss, explosion, or sabotage, would potentially result in the death of everyone within thirty to eighty miles from the inhalation of lethal levels of radiation, depending on the wind direction.[12]

In addition to the very real danger of meltdown in one of Hanford's production reactors, families downwind of Hanford lived within the pathway of radioactive fallout secretly discharged from the facility over more than four decades. Well before Hanford was operational, it was known that the chemical separation of plutonium at the facility would release radioactive and other highly toxic gases.[13] With the start-up of plutonium production in late December 1944, children living downwind of Hanford throughout eastern Washington, Idaho, western Montana, northern Oregon, and in communities downriver of the facility along the Columbia River became some of the first and youngest American civilians exposed to the ionizing radiation released by the Manhattan Project.

The First Civilian Victims of the Manhattan Project

Charlotte Rae McCormick, Marlene Campbell, Jay Mullen, Sally Albers Stearns, and Rance Jones were infants or young children when Hanford, then a wartime nuclear weapons production facility of the Manhattan Project, released its first, and some of its highest, levels of radioactive fallout downwind and to the waters of the Columbia River.

Here are their stories.[14]

Plaintiff 1: Charlotte Rae McCormick

As told by her son, Bob McCormick

My mother, Charlotte Rae McCormick, was born in 1938 in Emmett, Idaho. She had four older siblings and one younger cousin the family had adopted as a baby.

When Mom was very young, her family moved to Boardman, Oregon, a small farming community along the Columbia River. Boardman is located southwest and downriver of the Hanford nuclear facility. Mom's family lived so close to the river that its waters were visible from their house. My grandparents had sheep, cows, and horses. They also raised crops, including alfalfa. The family got their milk straight from the family cow and ate local produce, all of which were contaminated with radioactive stuff

Photo of Charlotte Rae McCormick as a child in Boardman, Oregon. Photo courtesy of McCormick family.

that drifted downwind and downriver from the Hanford nuclear plant. We were never told what my family was actually exposed to. All we were given was our estimated exposure dose to I-131. What about all the other stuff in the air and water that could have hurt us?

My maternal grandmother, well known and loved in Boardman, was part owner of a restaurant there. My maternal grandfather held many kinds of jobs, including road building throughout the Pacific Northwest, farming, and ranching. He also worked on the McNary Dam on the Columbia along with his two sons, son-in-law, and brother. Work on the dam began in 1947, bringing many jobs to the area. The dam was located about fourteen miles upriver from Boardman and was dedicated by President Eisenhower in September 1954. People working on the dam were unaware they were being exposed to contaminated air and water from Hanford.

My parents were married in Boardman and had a house right next to the river. My three older sisters and I were born in Hermiston, Oregon, but we grew up in Boardman. My sisters, from oldest to youngest, are Arla Renee, Tina Marie, and Robi Lynne.

We spent a lot of time in and on the river. My dad had a boat, and we spent every weekend we could swimming, boating, and water-skiing.

My family moved to Portland around 1968, when I was five. Several years later, my parents divorced, though my mom eventually remarried.

In 1988, when she was fifty years old, my mom began to suffer from swollen lymph nodes, fevers, itching, fatigue, coughing, and headaches. She consulted doctors to find out what was wrong. The diagnosis was large [histocytic] malignant lymphoma, a type of non-Hodgkin's lymphoma, combined with autoimmune hypothyroidism. We were all extremely worried.

Mom went through radiation and chemo for the lymphoma, which seemed to help her somewhat. Five years later, my stepdad died suddenly of a heart attack, a huge shock to my mom. Six weeks after that, my mom began to experience intense fatigue. She bruised easily, had night sweats and fevers, and started to suffer from headaches. She decided she had better have her doctor check whether the lymphoma was recurring.

Blood tests revealed her white blood cell count was off. Tests indicated it was acute myeloid leukemia rather than lymphoma this time around. Leukemia is a cancer of the blood cells. Both lymphoma and leukemia result from problems in white blood cells.

My mom was hospitalized for a full month to undergo chemotherapy and other treatment. Her two brothers came in three times a week to donate platelets to her. This was a big sacrifice for both of them, but they never complained. One of my uncles was a struggling farmer in eastern Oregon. He drove a long distance each time to help my mom. My other uncle made the trip by Greyhound bus from the coast. We were desperate, worried we were going to lose her. The platelet donations worked for a while. It was so hard to watch her suffer the side effects of chemo.

No one tells you about the sores, the explosive diarrhea, and the dehydration.

I stayed with her in the hospital, doing whatever I could, for the entire month. One of my sisters was sick at the time and could only come as far as the doorway of my mother's hospital room, wearing a mask, because Mom's immune system was destroyed by the chemo. My sister felt helpless. My other sister helped transport our uncle from the bus depot and back each time for the platelet donations. Mom responded well to the treatment and was discharged from the hospital in late July or early August 1993.

Early in December 1993, she developed an infection and began getting cold sores. The doctor, concerned, hospitalized her and prescribed a broad-spectrum antibiotic. For a while, she did a little better and was discharged from the hospital. In mid-December, she began to volunteer with the "Keep Christ in Christmas" program, as she had every year when my stepdad was alive. Volunteering seemed to keep her going. She was still very susceptible to infections due to the earlier chemotherapy.

After a few days, she started to get worse, becoming weak and increasingly exhausted. Her health declined pretty fast, and she was readmitted to the hospital. This time, desperate to help her, her oncologist pulled out all the stops, trying an aggressive experimental leukemia medication.

Nothing worked. We were devastated. We knew we were losing her. She lapsed into a coma, passing away December 18, 1993, at 4:20 p.m. I will always remember the date and time she died. She was only fifty-five.

My stepdad and mother died within six months of each other. My sisters and I didn't have time to prepare for their loss. My stepdad had just retired, and he and my mom had big plans to travel. They had saved for retirement, had just remodeled their house, and now they wouldn't be able to enjoy any of it.

Our mother's death devastated my sisters and me. Our mom was our rock. She had kept us together, and once she passed away, we began to drift apart.

I had heard something about Hanford on the news and began to attend meetings on Hanford to inform myself because I thought my mom's death might be related to her exposures as a kid to Hanford radiation. Right after

Photo of Bob McCormick, Charlotte McCormick's son. Photo courtesy of McCormick family.

she died, I contacted KOIN Channel 6 News in Portland. They did a story on her death, mentioning the possible connection to Hanford.

I began to put the picture together. Many members of my family had suffered from cancer or had died young. My maternal grandfather was the first, dying of lung cancer in 1961. My maternal grandmother developed uterine cancer. My maternal grandparents, my mother, and her siblings had lived in Boardman since just before Hanford began operations in late 1944 and started releasing high levels of radiation into the air and into the waters of the river.

One of my mother's sisters died from lupus when she was only fifty years old. Although I don't know what caused her lupus, I have read that lupus is known to be associated with radiation (uranium) exposure.

My mother's oldest brother, who grew up in Boardman and then worked at the McNary Dam, died of stomach cancer. Then, my mom passed away from acute myeloid leukemia after developing lymphoma.

The four of us and our kids have also had our share of health issues. My oldest sister had the longest exposure from Hanford. She has been diagnosed with a chronic precancerous gynecological condition, anal canal cancer that required a permanent colostomy, trigeminal neuralgia, severe neuropathy, severe lymphedema, fibromyalgia, depression, and anxiety. My second-oldest sister went through a hysterectomy due to uterine fibroids. One of her two daughters was diagnosed with acute myelogenous leukemia at age twenty-nine and is now in remission.

I had a heart attack in 2006, and I have ulcerative colitis, blood pressure problems, diabetes, and some kind of undiagnosed light-headedness. I don't know whether any of these health issues relate to our time in Boardman, but we were there during Hanford [plutonium] production years. I also wonder whether any of our health issues originated from our mother's exposures as a kid.

My family has suffered a lot of death and illness. As both the child of a Hanford downwinder and a downwinder myself, I would like people to know about the tragedy that has befallen all of us. Hanford operations greatly affected the people of the Pacific Northwest. The government did nothing to ensure that Hanford operated safely, without harm to downwind and downriver communities. There was no oversight of Hanford contractors and their safety procedures or lack thereof. Particularly after Nagasaki and Hiroshima, Hanford operators knew that releasing radiation downwind and downriver would cause harm.

It makes me angry that our government was involved in protecting the contractors throughout the Hanford downwinder litigation, making it impossible for people like us to get any accountability from those contractors. Indemnification agreements between Hanford contractors and the federal government benefited the contractors and caused detriment to the downwinders.

It is important that the sacrifices of the downwinders and their families

Civilians vs. Nuclear Workers

Federal law discriminates between civilian downwinders and nuclear workers who develop non-Hodgkin's lymphoma and leukemia after exposure to low-dose ionizing radiation. Non-Hodgkin's lymphoma and certain forms of leukemia are recognized as radiogenic and compensable under federal nuclear worker compensation law and under federal law compensating certain NTS downwinders. However, Hanford downwinders who develop these diseases are not currently eligible for compensation under federal law. The third disease mentioned in the McCormick story, lupus, has been recognized as potentially related to uranium exposure. See Chapter 10 for a discussion of the Radiation Exposure Compensation Act and the Energy Employees' Occupational Illness Compensation Program Act.

not be forgotten. We deserve, at a minimum, an official apology from the government that put our families in harm's way.

Plaintiff 2: Marlene Campbell

Marlene Campbell grew up about ninety miles north of Hanford in Ephrata, Washington. Marlene experienced frequent bloody noses and diarrhea during childhood and went on to develop severely disabling juvenile rheumatoid arthritis. Marlene believed that her childhood exposure to Hanford radiation was the cause of these health problems and the reproductive issues she faced later in life.

This is her story.

I was born on December 2, 1938, in Wenatchee, Washington. My family lived in Ephrata, about fifty miles from Wenatchee. Ephrata did not have a hospital, so my mother gave birth to me in Wenatchee. I lived in Ephrata until 1960.

My grandparents farmed about thirty miles from where we lived, between Moses Lake and Stratford. They came into town to sell free-range chicken eggs. All our fruits, vegetables, milk, and dairy products came from my grandparents' garden and cows.

Photo of Marlene Campbell as a child. Courtesy of Campbell family.

My parents worked when I was small, so I stayed every day with my grandparents on their farm. I was out in the field all day. In 1945, a year of high radiation releases from Hanford, I remember having bloody diarrhea and a bloody nose.

In 1945–1946, when I was in the first grade, I started feeling very sick. I never really felt well after that. I would cry morning and night because of severe leg and joint pains. The adults told me it was just "growing pains" and that no one wanted to hear about it any longer. In school, when I first learned to write with a pencil, my fingers hurt, and they were bent, crooked, and swollen. I was told that this was because I held my pencils and crayons too tightly. My eyes started to hurt, and I was unable to see the blackboard. This went on for years.

I had my tonsils out in the summer of 1946. A lot of other kids had their tonsils out too at the time. I don't know if it's normal to have so many kids having tonsillectomies.

When I was about eleven, I had red, warm, itchy rashes across much of my body. I was told this was "seven-year itch," whatever that is.

I remember a strange thing in the winter around 1950. My grandpa, uncle, my dad, and I went out to a pasture at my grandparents' farm and discovered nine of their Black Angus cows dead. It was really disturbing to see. No one knew why they had died. They thought it might be hoof-and-mouth disease. Their tummies were bloated, and they were all lying on their backs, their legs straight in the air. We also saw a lot of dead rabbits that winter—they were dropping dead everywhere along the roads. Everyone was told not to eat them—who would want to eat dead rabbits anyway? No one knew why the rabbits died. They just said the rabbits were diseased. This was different, though, from the other times when a disease came through and wiped out certain animals. That would happen sometimes, but when that happened, it would always take a year or two before the disease would end. This ended in six months. It was strange.

I was sick all the way through elementary and high school with rashes, bloody diarrhea, and bloody noses. When my periods began, they were irregular. No one knew what was causing these problems.

After high school, I went to the University of Puget Sound in Tacoma. I still felt bad all the time but kept plugging away. I could have gone further in school had I felt better. I suffered from sore muscles and joints, extreme

Marlene Campbell's hands: The disabling effects of juvenile rheumatoid arthritis, an autoimmune disorder. Photo courtesy of Campbell family.

fatigue, chronic sore throat, and flu-like symptoms with fever. These were arthritis symptoms, but I was told that I was too young to have arthritis, an "old person's" disease. Blood tests I had did not detect rheumatoid factor, a group of autoantibodies that appear in the blood of many people with rheumatoid arthritis.

Finally, in 1970, when I was in my early thirties, because of my swollen and deformed hands, a rheumatologist in Portland diagnosed rheumatoid arthritis. I later saw a specialist who diagnosed lupus, informing me that I had perhaps a year to live! I was devastated! Soon, thankfully, further testing revealed rheumatoid factor in my blood, meaning I had rheumatoid arthritis rather than lupus. I had been given my life back! In recent years, specialists have told me, due to the condition of my body, that I must have had juvenile rheumatoid arthritis.

I became a teacher and taught until 1974, when I had to go on disability due to all the serious physical problems I faced. Going on disability hurt me financially. Since I couldn't contribute the full amount to an IRA, I lost my retirement pension.

I didn't hear about Hanford's radiation releases until after the 1986 declassification of Hanford records by the DOE. It didn't take me but thirty seconds to realize that my exposure as a child to Hanford's airborne radiation was why I had been so sick and why I live in pain. I also immediately suspected that Hanford was the likely cause of the birth of a stillborn baby and the miscarriage I had suffered. I was never able to have children of my own, and I think Hanford was the cause of that. We adopted, and our daughter is wonderful, but not being a biological parent is a huge personal loss.

It just all hit me right then, when I heard about Hanford's radiation. I was devastated. I was kind of isolated down in Scottsdale, Arizona, where I lived at the time, away from all the news coverage at the time about Hanford. I immediately called my doctor. He said, "OK, we'll talk about it," but he didn't really seem to care.

Every blood relative I have is a downwinder, as they all had lived in Wenatchee. I tried to talk with them about Hanford's radiation releases and our possible exposures, but they didn't want to talk about it.

I didn't know what to do. The one organization that at least listened to me and took me seriously was the HHIN. I spoke with them and got some useful publications.

I also filled out one of the Hanford Individual Dose Assessment (HIDA) Project forms so I could get an estimate of my exposure to I-131. To get the estimate, I had to fill out a dietary form asking how many servings of milk and dairy products I consumed every day way back then. I had to guess on that. The first estimated HIDA dose for me came back low. I was very upset. I didn't believe it, and I told them I wanted them to recalculate the dose. They told me if they did that, someone else might not get their dose estimated. I told them I didn't care—that they owed me the recalculation, at a minimum!

So they went ahead and recalculated my dose. There was a difference between the two dose estimates! They were both presented within "uncertainty ranges" indicating the lowest and highest estimated dose in millirads I would have received.

The HIDA doses were based on the Hanford Environmental Dose Reconstruction (HEDR) Project. In the 1940s, my grandfather had a contract with the National Weather Service to track meteorological conditions out on the farm at Moses Lake, so I have total confidence that any weather data from our area that went into HEDR is accurate. My grandparents had a shed out there with equipment we weren't allowed to touch—it couldn't be jarred in any way. My grandmother and grandfather faithfully read and recorded the weather data off that equipment every morning and night.

In addition to being sick throughout childhood and developing juvenile onset rheumatoid arthritis, I have osteoporosis and Hashimoto's thyroiditis. I didn't know I had a thyroid problem until, as an adult, I asked to be tested. I have ongoing fatigue. I have horrible migraine headaches. I also have heart problems.

I am convinced that the arthritis, an autoimmune disease, was caused by childhood exposure to Hanford radiation. Over the years, the arthritis has caused extreme degeneration of my joints and hands. I have had both knees and hips replaced, along with my right shoulder joint. My left hand was so damaged that surgical reconstruction was unsuccessful. I have severe issues with degeneration in my elbows, ankles, jaw, and both hands. The arthritis has left me significantly disabled.

My husband is also a downwinder. In 1946, he moved to Soap Lake, about ninety-six miles north of Hanford. He now suffers from multiple sclerosis, an autoimmune disease. Even with this serious illness, he wasn't able to seek compensation or settlement for personal injury as a plaintiff in the Hanford downwinder litigation, as he didn't have thyroid issues. It was primarily the plaintiffs with thyroid problems and thyroid cancer who were offered settlement. On December 2–3, 1949, at the time of the Hanford Green Run experiment, he was out delivering newspapers, so he probably got a lot of exposure from that. He began to suffer from ongoing fatigue in 1950, intensifying as the years progressed.

The impact of Hanford operations on my life has been devastating. All these health issues have ruined my life. I try not to complain too much, but I live with significant disability. There is not a day that goes by when I am not in severe pain. I have multiple severe deformities from the rheumatoid arthritis. Osteoporosis has eaten the elbow in my left arm. When you are disabled, you live with it every day in every aspect of your life. You are no longer a part of the world.

Thyroid and parathyroid damage have also taken a toll on me. I have had abnormal serum calcium levels in the past, which can happen with parathyroid problems. I feel that either the hypothyroidism or the rheumatoid arthritis stunted my growth, as I am the only one in the family who is not around six feet tall. I'm five foot four.

I am disappointed in the medical community, as they didn't find anything wrong with me when I was young. I had significant health issues, yet no one took them seriously. No one suspected that these problems might be related to toxic or radioactive exposures. The health care community needs to be educated on the health damage radiation can cause. They could learn a lot from studying all of us. They should study the connection between our kind of "low-dose" chronic radiation exposure and autoimmune diseases, including rheumatoid arthritis.

I want those who read our stories to understand that Hanford operations caused grave physical and emotional harm. The care required to protect the public in the process of manufacturing nuclear weapons needs to be understood if we, as a country, are to continue to produce and stockpile these weapons.

I want the details of the Green Run experiment to be released. Many people in my generation, and most of the younger generation, have never heard of the Green Run. Often, they haven't even heard of the Hanford facility. That needs to be corrected.

We must also figure out how to protect the cleanup workers at Hanford. We must learn from our mistakes and put safety first. Scientists of the future need to understand the implications of what they are doing before they run amok and destroy the whole planet.

We need to speak out since we are the first generation of the atomic age. We need to make sure our experiences are documented—the good, the bad, and the ugly; everything.

The downwinders were not treated well. We were put in harm's way, and then our health damage was misunderstood and often disregarded.

Our attorneys were also victims in this. They were almost bankrupted representing us. They need to be recognized for their sacrifices.

We've lost some very important people in our generation because of Hanford—they never had a chance.

Marlene Campbell passed away on January 18, 2018.

Marlene Campbell. Photo courtesy of Campbell family.

Juvenile Rheumatoid Arthritis

Juvenile rheumatoid arthritis is often referred to today as "juvenile idiopathic arthritis" (JIA). JIA is an autoimmune disorder in which the immune system attacks the body's cells and tissues as if they were foreign bodies. This type of arthritis is defined as joint inflammation leading to pain for a period greater than six weeks in a child sixteen years of age or younger. In about one-third of patients, JIA continues into adult life.[15]

Symptoms can include fever that comes and goes, rashes on the child's arms and legs, anemia, and weight loss. Some children also have growth issues that impact overall height. Eye problems, including iritis or uveitis (inflammation of the eyes), can occur, causing blurry vision. JIA can be progressive, and current treatment recommendations include regular screening through an ophthalmologist. Children are sometimes diagnosed late with JIA because they may report vague symptoms or symptoms may be misinterpreted as "growing pains."[16] Due to limitations on physical activity relating to the condition, children with JIA often feel isolated from their peers. Permanent damage from JIA is rare today, whereas decades ago, due to limited understanding of the condition in combination with limited treatment options, JIA could result in significant, lasting disability.

The precise cause of JIA remains unknown, but autoimmune disorders can develop as the result of toxic exposures during fetal and infant life. Research into the environmental triggers of JIA has lagged behind research into the environmental triggers of other autoimmune disorders, but researchers believe that a genetic factor causes increased susceptibility to JIA when an environmental trigger occurs. Epidemiological studies of atomic bomb survivors have also determined that DNA damage from exposure to ionizing radiation can lead to autoimmune diseases, including rheumatoid arthritis.[17]

Multiple Sclerosis Following Exposure to Low-Dose Ionizing Radiation

Multiple sclerosis (MS) is an unpredictable, disabling autoimmune disease that targets the central nervous system (CNS). With MS, a

continued

continued from previous page

person's immune system attacks the myelin sheath of fiber that protects the tissue of the CNS. Symptoms of MS typically include slurred speech, impaired coordination, and other neurological symptoms, including blindness, numbness, and dementia.

Studies have determined that there is a potential link between MS and exposure to ionizing radiation.[18] In 1983, an epidemic of MS occurred in Spokane County, Washington. Researchers hypothesized that there might be a connection between this epidemic and the "contamination of eastern Washington from a very heavy 40 year downwind radiological fall-out from nearby US Hanford Nuclear Reservation nuclear weapon production."[19] Fallout from US and Soviet atomic tests had been deposited in the area as well. Furthermore, two controlled studies of MS in Sweden reported that people exposed on the job to ionizing radiation had over a fourfold higher risk of MS.[20]

Abnormal Nosebleeds

Does radiation exposure cause nosebleeds? Reliable statistics are hard to come by, but anecdotal evidence is recent and strong. Doctors treating those within the exposure areas of the March 11, 2011, catastrophic reactor meltdown at the Fukushima Daiichi Nuclear Power Plant in Japan began to see "increased nosebleeds, stubborn cases of diarrhea, and flu-like symptoms in children."[21]

Abnormal nosebleeds were also reported following the Chernobyl nuclear disaster in 1986, and data from over two thousand people who evacuated from Pripyat eight to nine days after the accident reveals a 21.6 percent incidence of unusual nosebleeds. Dr. Hida Shuntaro, a medical doctor who was also a survivor of the atomic bombing of Hiroshima, observed abnormal nosebleeds in his fellow survivors.[22]

Researchers theorize that exposure to low-dose radiation involves minute airborne radioactive particles that can enter the nose and stick to the surface of the capillaries. The particles could emit alpha, beta, or gamma radiation, depending on the radioisotope involved.

The radiation could directly damage the membrane of the capillary. Another theory is that hydroxyl free radicals produced by the radiation cause the damage, which could then lead to nosebleeds.[23]

The controversy over abnormal nosebleeds and radiation exposure revolves around whether exposure has been high enough to cause these nosebleeds. Exposures at Fukushima, Hanford, and Chernobyl (other than to liquidators and others involved in trying to contain the disaster) are termed chronic "low-dose" exposures.[24] The term "low-dose" can be rather confusing, however, as even moderate or high doses over several years can be seen as a cumulative dose from a series of multiple low doses.[25]

Plaintiff 3: Jay Mullen

Jay Mullen was born in 1940. From late 1944 through 1945, when Jay was still very young, he lived with his family in Idaho at Farragut Naval Station, southeast of Sandpoint on Lake Pend Oreille, where his father was stationed as a medical corpsman during the war. During this period, some of Hanford's highest releases of airborne radiation, including I-131, occurred. Documents declassified by the DOE in 1986 revealed that Hanford's airborne radiation was deposited within a broad geographic region, including northern Oregon, eastern Washington, Idaho, western Montana, and southern British Columbia.[26] Farragut Naval Station was within the Hanford downwind exposure area. When he was almost nineteen years old, Jay's thyroid began to act up, and the end result was paralysis.

This is his story.

I was born January 22, 1940, in southeast Missouri, in the little town of Cape Girardeau on the banks of the Mississippi River.

I spent my childhood in Cape Girardeau except for a hiatus during World War II during which my father enlisted in the military. When I was four years old, he was stationed in Idaho at Farragut Naval Station, southeast of Sandpoint on Lake Pend Oreille. Farragut was the second-largest naval station in America during the war. It was a processing point. My dad was a medical corpsman. There was a hospital at Farragut, so first he

Jay Mullen and his sister. Photo courtesy of Mullen family.

attended to the wounded in the hospital, then he was rotated out into the Pacific theater.

I was there at Farragut for Christmas of 1944 and into 1945, when some of the highest levels of radiation were released from Hanford. It's been well documented that the radioactive fallout from Hanford reached Idaho.

I attended a day care center while my mother worked. They made us drink milk; milk that, we now know, contained radioactive iodine. Fallout from Hanford drifted over the area, contaminating the pasture grass. Cows grazed on that grass and took in the radioiodine, which went into their milk.

I had this thing for ice cream at the time, which likely gave me even further exposure. Just to illustrate how strong that attraction was, I vividly remember a daring adventure I undertook as a five-year-old on St. Patrick's Day, 1945. They were giving away ice cream in one of those Quonset huts where they showed movies, so I snuck out of the house all by myself and went over and slipped into the event.

I became a sort of little mascot for the military guys, and they even made a uniform for me. I would march in Friday parades in that uniform, a miniature sailor suit. There was a soda fountain at the base, and sailors heading overseas would put me up on the counter and buy me milkshakes and ice-cream cones. So I got more than my share of dairy products on a regular basis. I also drank copious quantities of milk. I was ingesting Hanford's radioactive iodine big-time that way, and during the part of life I was most vulnerable to radiation.

My sister and I had moved to the naval station with my parents. My older brother stayed back in Cape Girardeau with relatives. After the war, in 1947, another brother was born back in Cape Girardeau. Only the two of us who lived in Idaho at the naval station developed thyroid problems. My thyroid issues were severe. My sister, five years older than me, had mild hypothyroidism. My older brother, with the same genetic makeup but who stayed behind in Cape Girardeau, had no thyroid problems, and neither did the brother born after the war.

My thyroid began to go haywire when I was a sophomore at the University of Oregon. My father had so enjoyed the West that after the war, when I was around fifteen, he decided we would move back. I was then a junior in high school. We moved to Medford, Oregon. I enrolled at the University of Oregon in the fall of 1957, after graduation from high school.

In the fall of 1959, just before my nineteenth birthday, my thyroid just started going berserk. I had been feeling really sick; my hands were shaking, and I was sweating profusely. I went in to the student health service, and the doctor looked at me, shaking and sweating, and told me to come back in the following week for a basal metabolic rate test. I was scheduled to go in the following Tuesday, but that weekend I wasn't feeling well at all, so I went and checked myself into the infirmary.

My parents and sister came up to bring me a birthday cake and found me there in the infirmary. My parents spoke with the doctor, who told them something was wrong with my thyroid. He suggested that they take me out of school to address the issue. So they did. I spent a few days in the hospital in Medford being observed; then I came home. The next day, I woke up and found I couldn't move! I was paralyzed from the neck down. I'd been a university athlete; now here I was, paralyzed.

They didn't have a clue what was causing this. Somehow or other they attributed it to the thyroid. When I first got sick, the symptoms looked like hyperthyroidism. Then the paralysis hit, and they couldn't figure out what was going on. The doctors in Medford first thought it was psychosomatic, all in my head.

The doctors in Medford couldn't figure out what was causing the paralysis, so they called the University of California, San Francisco Medical Center (UCSF) to set up a consultation. I went down there, and the UCSF metabolism clinic thought my medical issues were so freakish that they asked my father if he would bring me back later so they could study me. The UCSF doctors were a bit more sophisticated than those in Medford. They found they could induce paralysis by changing my serum potassium levels. They did hundreds of tests, spinal taps and so forth.

They said I was only the eleventh person in medical history with these weird medical issues. The paralysis was intermittent. I got to where I could feel which muscle was going to go next and when it was coming on. They concluded that my thyroid had to be removed, so I had a total thyroidectomy. The thyroidectomy cured the paralysis.

Unfortunately, while they were operating on me, they found that the thyroid gland was embedded in a neck muscle, so it was a long and complicated surgery. They pinched a nerve in my shoulder during the procedure, so my left arm was paralyzed when I woke up. The paralysis in my arm lasted about six months.

I returned to UCSF some time later, and they studied me without the thyroid and without the paralysis, so they had the before and after. The first time I was seen at UCSF, the costs were high, nearly bankrupting my family. The second visit they paid for, as I was then part of a research study.

After I had recovered to the point that I was in good health, I went back to live with relatives in Cape Girardeau because the hospitalization had run us financially into the ground, and I could go to school inexpensively back in Cape Girardeau. I later went back to the University of Oregon and finished up there with a degree in history, followed by a master's degree. Then a doctorate from the University of Kentucky in 1971.

The story on Hanford is that there was this insidious presence, or force, constantly being released into the environment. It did in fact impact people's health, but every time this has been even remotely suggested, it has always been pooh-poohed and swept under the rug. I believe the downwinders were the canaries in the mine.

I first connected myself to Hanford following the great release of documents by the Department of Energy in 1986. And I had a student at Southern Oregon University, where I taught, who was an antinuclear activist at Diablo Canyon. She had been arrested down there protesting. She is the one who connected the dots and told me that Hanford's radioactive iodine was likely the cause of my thyroid-induced paralysis.

Once the radiation discharges from Hanford were made public, I thought I would tell "them" what happened to me and that they would be so interested, in their desire to protect the public, that this information would be disseminated! I wrote a letter to the Centers for Disease Control and Prevention (CDC), telling them that I had lived downwind at the Farragut Naval Station in Idaho, that I was paralyzed, and that I went through this whole experience. I got back a supercilious letter saying they weren't interested in my story, that I wasn't "part of good epidemiology," whatever that means.

I thought that, certainly, once Hanford's releases were common knowledge, the government would want to start turning over all the stones to protect the public. And what I found out is that they don't want the stones turned over; in fact, they want other stones heaped on them to obscure them.

Our family once had a large trunkful of letters my father had written, but only one letter survived. That was the one commenting about dropping the atomic bomb. My dad thought the bomb saved him from having to patch up the wounded from the invasion and perhaps even saved his own life. The irony is, while he thought he was defending his family out there as part of the military, his family back at Farragut, downwind of Hanford, was being irradiated by the very thing he thought was defending him.

There are three famous things about which I felt guilty that happened during World War II to kids my age. First, I saw a picture of Douglas MacArthur's son, who is a year older than I am, at the Battle of Corregidor. He was standing there with his army hat on. I identify that with my childhood

Jay Mullen. Photo courtesy of Mullen family.

sailor suit. I thought to myself, "Poor old Arthur MacArthur was over there with the rockets' red glare and bombs bursting in air." Then I remember seeing that photograph of a little Jewish boy in Poland holding his hands up while a German soldier pointed his rifle at him. I thought, "Look at that little boy; he was out there in harm's way, and there I was running around in my little sailor suit." And then, I read a book way back when called *Three Came Home* about the son of a Berkeley woman who had married a New Zealand woman. They were living in Indonesia when the Japanese took it over, and they were put into a camp. He was my age. So here were these three guys, and I thought, "Boy, those guys had it rough." And the irony is, I was the one who was damaged by the goddamned bomb, and the war made me a bigger victim than the three of them. I don't know, maybe that boy in Poland got killed, I don't know. I often thought, "If the only thing I can do is tell a story, then that's a pretty vivid comparison."

I want people who read our stories to understand that on the one hand, there is a danger out there, and on the other hand, the people who should be interested in the welfare of the public are more interested in the welfare of the technology and the nuclear industry.

The only thing I hope can come from this is that the song can be sung so that the people can hear it.

Jay Mullen passed away in July 2016 following a heart attack.

Plaintiff 4: Sally Albers Stearns

Sally Albers Stearns came to Richland in 1944 as a two-year-old, living with her family in the Desert Inn Hotel on George Washington Way until construction was completed on the family's "A house." She attributes her father's death from cancer, and her mother's health issues and her own health problems to radiation exposure from Hanford.

This is her story.

I was born in Denver in 1942 at the Salvation Army hospital for unwed mothers. I was adopted shortly thereafter and raised as an only child. I think of my adoptive parents as my true mom and dad.

My dad worked at the time for DuPont in Denver, Colorado. One day, someone from the FBI came and informed him he was to go to Richland, Washington. My dad hadn't been drafted into the military because the night before he went in for his physical, my mother cooked up something with a lot of sugar and butter, and so he didn't pass the test. He was tending toward diabetes at the time, and the sugar and butter made his blood glucose levels climb perilously high. That is why he ended up at Hanford during the war.

So first my dad went to Richland for a year without us. I think this was because there was nowhere there for families to live at the time except barracks and tents in North Richland. The town of Richland, with its alphabet

Sally Stearns and her father. Photo courtesy of Stearns family.

houses, was still being built. He worked in various areas out at Hanford, although I'm not totally sure what he did. He never talked to us about his work. There was a code of secrecy at the time, and no one was allowed to speak about their jobs. I later learned he was a chief instruments specialist at Hanford.

My first "home" in Richland in 1944 was the Desert Inn Hotel on George Washington Way, where we lived until our house was ready. We finally moved into our alphabet house, an A house, in Richland. Two families lived there with the stairway in between, so it wasn't too noisy.

On Saturdays, when I got a little older, we kids would all go to the Uptown Theatre in Uptown Mall on George Washington Way to watch the movie and get candy for ten cents. We would still have time to go to have Spudnuts, which are doughnuts made with potato flour. They are delicious! The following day, it was our "job" to go roller skating. Weekends with my friends in Richland were good. Richland was a safe community for kids—at least on the surface!

I remember when we bought a car. My dad bought a Studebaker, and he was really proud of it. I can see it parked out front of our A house in some of the childhood pictures of me taken in Richland. I was very close to my dad. I told him everything. I remember that my dad and I used to compete in three-legged races in Richland during some kind of family festival they held there. Once he twisted his ankle during the race. I also remember doing tap dancing with him at the community center. Part of the building still stands. We did a lot together.

On the other hand, my mother was rather distant, and I didn't tell her much at all until I had grown up and had a family of my own. Then, communications improved. I think she was really afraid of bonding with me because she worried that my biological mother might come along at any time and take me away. I honestly didn't have a very happy childhood. I think it was because I was adopted. It caused a lot of issues for my mother, and I was always afraid of being alone.

My parents both smoked several packs of cigarettes a day. All the adults and the teenagers smoked back then. It seemed like everyone had a cloud of smoke rising over their heads.

In Richland, I remember that every morning my dad would have to go down in our basement to shovel coal. They delivered the coal from a big dump truck through a chute that led to the basement. I recall the sound of the coal truck pulling up to the side of our house and then the crashing and clanking of the lumps of coal as they cascaded into the bin. The house was freezing on winter mornings until our coal furnace finally started to heat it up.

I also clearly remember the sound of bottles clanking and watching a little truck coming down our street. I thought it was the milkman bringing milk, but the driver was on an entirely different mission. He would stop at each house to pick up the bottles containing urine samples from our dads. I think they collected urine samples almost every day, to monitor internal radiation levels from their jobs, I guess.

When I was a kid, I would wake up in the middle of the night with blood all over the front of my nightgown from a major nosebleed. That happened a lot. Each time, I had to be taken to Kadlec Hospital because no one could stop the nosebleeds. Each time, they would have to pack my nostrils. I have heard that nosebleeds can be symptoms of radiation exposure—I know Hanford released most of its airborne radiation at night.

During the polio epidemic in the early 1950s, my mom kept me isolated. I always got things much worse than everyone else, so she was probably afraid I would be highly susceptible to the polio virus. I'm actually glad she isolated me. I think my immune system was weakened from all the radiation exposure I was getting there in Richland.

I had really bad allergies. I was one of the first kids in Richland to have skin allergy tests. They were really uncomfortable. The allergies got to the point where sometimes I couldn't breathe. One day, the boy next to me in school had been riding horses, and when he sat down next to me, I suffered a major allergic reaction. I remember waking up at Kadlec Hospital with oxygen.

I had a dog named Mumpsie Mary, who you can see in one of my photos from my time in Richland. I didn't take care of her well. My mom finally got fed up and gave her away one day when I was at school. I was *so* mad. My mother would sometimes say and do really mean things.

I was in a Richland Brownies troop, but I never went swimming with the troop because I was overweight, and I didn't want to be made fun of. There were a lot of things I didn't do as a kid because I was self-conscious due to my weight. I think that the weight I put on as a kid was from hypothyroidism, even though I wasn't diagnosed with hypothyroidism until decades later, after the DOE released its documents on Hanford in 1986 and admitted that radioiodine and other radiation were released into the community. Then everyone got checked for thyroid disease. I was always lethargic as a kid. I think that was from the hypothyroidism too.

Mother would see a house she wanted outside my current school district, and we would move and then move again. I went to five different schools. One of those was Jason Lee Elementary School. I remember there were big metal things, sort of like movable walls, in the hall that would lock around us, to protect us, I think, during the many air-raid drills they held at the time. They slid out from the walls in the hallways, enclosing us in cubicles. All that caused me to be claustrophobic, which I still am.

In school, we also had a lot of "duck-and-cover" drills starting around 1951 because the Soviet Union had detonated its first atomic bomb in 1949, making everyone nervous. They were worried about the "Reds" [Russians] launching a nuclear attack on Hanford. During these drills, I remember that the sirens would go off, and they were terribly loud. We would dive under our desks and put our hands behind our heads. I'm not sure what good that did. The sirens made a wailing sound, and I hated it.

In case of an attack on Hanford, if kids were in school, we were supposed to be bused out of the area. Just a few years ago, when a military plane flew very low over our house and started to turn, I began to shake, as I was afraid we were being attacked, because we still live close to Hanford! This gut reaction must have been left over from what I went through in childhood in the Tri-Cities.

As a teenager, when my periods started, I had horrible bleeding. I ended up in bed each time for five days. I had to have periodic transfusions when I was young due to loss of blood, although I don't know if the blood loss was related to my periods. I also ended up with problems with my reproductive system. I have had lots of uterine tumors and fibroids. The last one was so huge that it engulfed my ovary. The doctors thought it might be cancerous, so they removed it.

I have arthritis and a lot of digestive problems as well. I try to overlook many of these and the other health problems so as not to be depressed. The joints in my thumbs and other places have disintegrated. I also have fibromyalgia-like symptoms.

I am always tired, but now with vitamin B12, I'm better. I was sick for a year and a half until my doctors figured out I had a staph infection in my gut. It was so bad I thought I was going to die. I started to give away things to prepare for my impending death. It was that bad.

My mother also had health problems caused by her time living near Hanford. I'm convinced of that. I think she was probably hypothyroid, as she was overweight. She always looked like she was nine months pregnant.

She had several miscarriages, was never able to carry a child, and had serious diabetes. She was always in the hospital with horrible boils, and they would put Denver mud [also called antiphlogistine] on them.

My mother died before my father. I wish I could have had my father around longer. I really miss him. He died from a disease called polycythemia vera, which is a slow-growing type of blood cancer in which the bone marrow makes too many red blood cells. This causes clotting and strokes. I know he got a lot of exposure on the job at Hanford, and I believe this was the cause of his death from blood cancer. I filed a claim on his behalf under the federal nuclear workers' compensation law. I was really unhappy to discover that it was extremely hard to get compensation, even though my dad's cancer was on the list of radiogenic cancers recognized under that law. For that reason, I haven't gone on the tour of the B reactor out there at Hanford. I guess it's my way of expressing my displeasure. My dad gave his life for the government's nuclear weapons program, and then the government didn't want to provide compensation for his death. It took me four years to finally succeed in seeking compensation for his death as a nuclear worker. That makes me angry.

I have an official document acknowledging the importance of my father's contribution to the Manhattan Project. It was signed by Secretary of War Henry Stimson in 1945. My dad didn't say anything about it while he was alive. I found it among his papers after he passed away. It is an amazing historic document. You'd think that the people in charge of compensation of nuclear workers would have acknowledged his contribution like Secretary Stimson did!

I have a lot of thoughts about what the children of Hanford have gone through. I feel in my heart that we children are under a microscope. "They" watch our lives as we go on. "They" watch our health issues. I know a lot of kids I grew up with who now have lupus, hypothyroidism, cancers, and other diseases. "They" keep track of all of that, whoever they are.

I know "they" are watching us, because when my dad died, and I filed the claim under the nuclear workers' program, I found his two obituaries carefully stored in his file box with his exposure records at Kadlec Hospital. Someone put them there along with his exposure records. Someone is keeping track of all of us, workers and families.

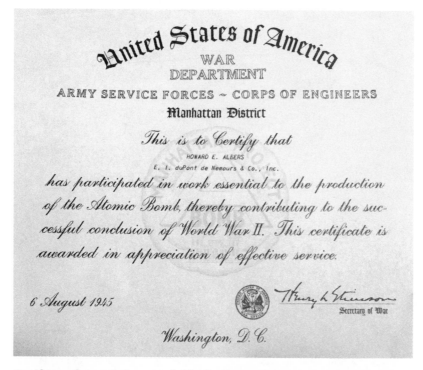

Certificate of appreciation awarded by the US War Department to Sally Stearns's father for his participation in producing the atomic bomb at Hanford, August 6, 1945. Photo by author.

I saw some photos of my dad's coworkers when I visited the museum in the Tri-Cities. I think it was probably the Columbia River Exhibition of History, Science, and Technology museum in Richland. But, there was no mention of those of us who grew up here being sick or even dead now due to our exposures as kids. When I look at photos of kids I knew, half of them are dead now. Aren't we part of the story?

As to the Hanford downwinder litigation, what I am tickled about is that by settling with us, the contractors and the government more or less had to admit they did this to us, at least regarding the thyroid cancer and disease. I didn't care about that little bit of money I got in settlement that I could go to the store and back with. The people in Hiroshima and Nagasaki were probably affected the same way we were.

Sally Stearns. Photo courtesy of Stearns family.

I wish people cared more about those of us whose lives have been sacrificed for our country in this way.

Plaintiff 5: Rance Jones

Rance Jones was born in 1945 at the Atomic Energy Commission's Kadlec Hospital in Richland. As an adult, he worked in several capacities at the Hanford facility. Today,

Hanford's Doorstep Urine Collection Program

Plutonium that has been in the body for more than two weeks is excreted very slowly in the urine. The amount of plutonium voided in urine was used to estimate Hanford workers' plutonium body burden. The presence of other fission products, including I-131, barium-140, strontium-89 and -90, some rare earth elements, yttrium, and certain beta-particle emitters was also detectable through urine collection and analysis.[27]

Monitoring of worker exposures was originally conducted by the Health Instrument Division of the Hanford Medical Department.[28] In order to avoid picking up contamination at the workplace, "the Medical Department . . . proposed that the samples be taken in the employee's home so that he will normally have removed his work clothes and have bathed before the sample is to be taken."[29] Samples were "obtained by delivering equipment to the individual's home and collecting all urine passed thirty minutes before retiring and thirty minutes after arising for two consecutive days."[30] The sampling period for day workers was Saturday and Sunday, and for shift workers, it was Wednesdays and Thursdays that they did not work.[31] If plutonium was found in the sample, five morning and evening samples were obtained for recheck.

Hanford medical officials worried that the home urine collection program would cause public concern. "As for Security, it will be impossible to operate the program without exciting some comment in the home of the individual. Employee cooperation in treating the test as a normal routine will help to eliminate discussion of the procedure off the Plant."[32]

he suffers from multiple cancers that he feels were caused by his radiation exposure in Richland as a fetus, infant, and child in combination with his exposure on the job at Hanford. Jones was a participant in the Hanford Thyroid Disease Study (HTDS). Because he wasn't diagnosed with thyroid cancer until after the study was concluded, he was reported as an HTDS participant without thyroid cancer.

Here is his story.

I was born in 1945 at Kadlec Hospital in Richland, Washington. At that time, Kadlec was a government-owned hospital, like everything else in Richland.

I know now that I was a twin, but I didn't find that out until I was thirty-five years old. There were a lot of things I only learned long after the fact. For instance, in 1944, my mother, pregnant with me, volunteered at Hanford for a prenatal study. They told her they were giving her prenatal vitamins that would help her future children's teeth. I would later learn that this study was part of Project Sunshine, conducted under the auspices of the Manhattan Project, in which the dead babies' and children's bodies and teeth were collected to study the effects of strontium-90 from nuclear fallout on growing bones.

At eight and a half months' gestation, any record of a twin disappeared. It had heart tones up until eight and a half months, but after that time, there were no more comments regarding my twin within my mother's birth records.

Hanford was considered a "top secret" facility. Back then there was a "don't tell, don't ask" policy for everyone who lived in Richland. You wouldn't think a pregnant woman would fail to ask about a child who had suddenly disappeared, but back then, during childbirth, they would knock a woman out with ether, and the father wasn't allowed in the delivery room. So neither my mother nor my father knew what had happened. They were given a baby to take home. That was it.

I grew up in a B house on Barth Avenue in Richland. As a child, I was treated for allergies, asthma, and hypothyroidism. Around 1963, when I was eighteen, I was working up north in a farming area. I developed bacterial spinal meningitis, an infection of the spinal column. I was in a semicoma, my body contracted into the form of a C. The doctor came out on a house call, quarantined my house, and told my folks that I probably would not live twenty-four hours. They hauled me in to Kadlec Hospital. Somehow, I didn't die. They kept me in Kadlec for a week, trying to figure out why I hadn't succumbed to the spinal meningitis. I survived, in my opinion, because I am Mormon, although at the time, being eighteen years old, I wasn't acting much like a Mormon. When I got sick and was in the hospital, my father and another Mormon man gave me a religious "priesthood blessing," and I remember that I immediately felt better.

Rance Jones as a child. Photo courtesy of Jones family.

Around the same time I was diagnosed, two military people from Fort Myers in Tacoma died of spinal meningitis, so there was definitely something going around. My tonsils also essentially exploded, necessitating an emergency tonsillectomy.

One of the complications of having spinal meningitis was that I lost a lot of nerve endings, and now, as a result, I can feel pressure, but I can't feel pain. I have to watch myself. I can stick my hand in a fire and not feel it. When I was a young father, I was holding a bunch of unlit sparklers, and they ignited and welded to my hand, but I couldn't feel it.

From 1978 to 1979, I worked as a union carpenter/union steward, helping build the Hanford reactors for the Washington Public Power Supply System (WPPSS) and the Fast Flux Test Facility (FFTF) reactors. I also worked adjacent to the reactor core during a refueling outage at the WPPSS reactor in 100 N Area.

From 1980 until 1995, I managed railroad operations for the Department of Energy at Hanford, overseeing highly radioactive shipments and radioactive excess equipment burials via rail cars. The government had a railroad there, and I had a background working on railroads, so they hired me as manager. We moved the radioactive materials out of reactor areas and chemical separations areas, did the radioactive excess equipment burials, and moved the coal around the area for the power plants.

In 1980, when I first went to work at Hanford, I needed a certified birth certificate. At the time, I only had a copy of the hospital birth certificate required for my school records, not a certified birth certificate. When I filed for my security clearance, my certified birth certificate came back stamped "Boy baby, DECEASED AT BIRTH." So it looked like I was dead! I had four children at the time. I was definitely not dead. I went to my life insurance agent, who was a friend, and I gave him the birth certificate and told him I wanted to collect on my life insurance policy!

The FBI froze my security clearance approval for employment at Hanford for six months due to that "Boy baby, DECEASED AT BIRTH" stamp. It took a notarized statement from both the nurse who attended my birth, now in a nursing home, and the attending doctor, who was ninety-one, asserting that they specifically remembered my birth before I was able to get a new certified birth certificate. I took my copy of the "Boy baby, DECEASED AT BIRTH" certificate with my name on it and showed it to my mother, She broke down crying. She said, "I always wondered what happened to your twin!" She had lived with that question for thirty-five years, and they wouldn't give her any information.

About fifteen years later, I came across information on the AEC and Project Sunshine. In the article I read, they admitted to experimenting on fetuses and newborns that had died. I wondered whether my twin had died and then become part of Project Sunshine.

My twin may or may not have been born alive. I was able to track the fetus. I found out that on May 6, 1945, the day after my birth, a nitrogen canister was shipped from Kadlec Hospital via the AEC airport in Richland, which was a military airport, to an undisclosed site in Chicago. I believe that site was the University of Chicago, a special isolated lab of the Manhattan Project called Met Lab. There, they studied the effect of radiation, especially of strontium-90, on those little bodies that were shipped there. The parents were not told that their infants and fetuses were being snatched and studied. It was done without getting releases or telling the families about it. That must be why they were worried that the public would have concerns if they found out.

After I got married, I moved to Kennewick, but my folks lived continuously in Richland. My father worked at Hanford until he retired. His last job was at 100 N, where the reactors were located. He was a chief chemical operator for the project. After he retired, he went back to work as a security escort in 100 N. He had prostate cancer and six bypasses during his life. My mother was not in good health either. She had cancer several times and died of bone and breast cancer.

I remember the last time I ever talked to my father. I was at my job as a Hanford train dispatcher/manager, operating the radio while I ran railroad operations. I got a phone call from my father, who was working as a security escort in 100 N. He said, "Son, I'm not feeling well. Phone in to the nurse." I called the nurse. Later, over the radio, I heard the call come in from the ambulance transporting him to the hospital, and I was able to meet him to say good-bye at the hospital before he left us. About two hours after we got him into the hospital, he went into a coma, and three days later, he was gone. His pancreas basically blew up with pancreatic cancer. They didn't want to document pancreatic cancer, so they just said he "passed away." But his treating physician said he had cancer and that his pancreas was like Swiss cheese because of it. He was diagnosed with cancer two times, but his medical files no longer contain a record of that.

In 1990, while I was still working at Hanford, I had diverticulitis with an abscessed [blocked] colon. I was in the hospital several days. A friend of mine, a doctor, managed to open it up, clear the blockage, and treat it with IV antibiotics. He said if that hadn't worked, he would have had to cut the colon out.

I retired from Hanford in 1995 and left Richland and the Hanford area in 1999.

I have been diagnosed with nine cancers since March 2000.

I am convinced that these cancers are the result of exposure to environmental radiation as an infant and child in Richland, in combination with exposure during my employment at Hanford.

In 2000, the doctor found chondrosarcoma, or bone cancer. I had parts of three ribs removed on the right side and a metal mesh cage, about four by six inches, put in to replace the ribs and protect the lung. This has caused my right lung to be unable to expand. Since the lungs work in concert, the other lung couldn't fully expand either. I have chronic obstructive pulmonary disease (COPD) because of it. During the rib surgery, they also took a large cancerous tumor off my right lung. They told me they had gotten it all, and they sent me home.

It was extremely difficult to breathe. Ironically, due to the effects of the spinal meningitis that had dulled the pain sensation in my nerves, the day after I got out of the hospital, I was able to return to work. I had an office job, and I was able to sit in my office chair, which was tolerable as long as I didn't breathe very deeply. Breathing only shallow breaths, I've learned how to function.

I asked for an analysis of the biopsies because of my background at Hanford. The doctor at Sloan Kettering Hospital in New York City sent me a copy of my lab report. He probably didn't mean to send it to me. The isotope on my rib/bone cancer had been isolated and identified. It was like looking at a DNA analysis of a cell, and the DNA analysis identified cesium as the cause of my rib cancer. Since then, ironically, they have lost the lab report.

Fortunately, I was required to come in for a six-month follow-up, and during the follow-up, an MRI found cancer on the left side. This was 2001. They were going to take out another rib. When you donate three ribs, it's not fun. I wasn't looking forward to losing yet another one.

The second time they found cancer, luckily, I had learned from my research that there was a new way of treating radiation sickness and cancer. Not just through surgery and chemotherapy but by boosting immune system function through a mineral supplement. I found where the immune system mineral supplement was made, helped the company develop it, and they allowed me to buy the supplement. I was a divorced father of seven, and I was concerned that my kids would grow up without a father if the supplement didn't work.

I didn't believe at first that the mineral supplement would work. I talked to people who had taken it. I knew that the supplement would strengthen my immune system and that cancer was a disease involving the immune system. I took it for six weeks. This was in 2001. In six weeks, this mineral product boosted my immune system, eliminating the cancer on the rib. They had scheduled me for surgery for the cancer, and they had to do a presurgery MRI. The MRI revealed that the cancer was gone! The oncologist called me up and said, "How are you feeling?"

I told him I was just going to work, then going home to bed because I didn't have a lot of energy. I asked, "When are we doing the surgery on the rib on the other side?"

He said, "Well, I have good news." He then asked, "What have you changed; what are you doing?"

I had promised the manufacturers not to tell many people about the mineral supplement. So I told him I was just going to work, going home, and going to bed to rest; I didn't tell him about the supplement. My oncologist is also a Mormon, and I finally told him that one thing I did was to get a priesthood blessing.

He said, "Well, it must have worked, because your cancer is gone!"

Since then, I have done a lot of research on the mineral supplement. It's nothing more than an immune system builder that helps my immune system fight off the challenges of cancer.

In August 2001, the DOE conducted a beryllium study. Hanford used metal beryllium, and breathing in fine particles of the metal can cause beryllium sensitization, which can lead to lung damage and chronic beryllium disease. When they ran the beryllium study on me, they ran it not once but three times because they told me I couldn't have a beryllium reading as high as mine, or I would be dead.

Why did I have such a high beryllium reading? From 1978 through the latter part of 1979, I worked at Hanford as a carpenter and a carpenter steward, and part of my job was to help build nuclear reactors there. We built Plant 2 which is now called the Columbia Generating Station, and we also built Hanford 1 and 4, which they later tore down. Then we built the FFTF. I also worked adjacent to the reactor core during a refueling outage in the N reactor. I worked in Hanford Plant 2 in the containment area. Pipe fitters and boilermakers were installing the different supports and piping and doing a lot of grinding. I don't know if that is where I got the beryllium exposure or if I got it when I was a train supervisor working in or around all the facilities that used beryllium. We visited and supported all those facilities.

In approximately 1998, I filed both beryllium exposure and Hanford downwinder claims, as I was impacted by both. I was told I had to choose between the claims, so I withdrew from the beryllium study.

I was a participant in the Hanford Thyroid Disease Study. The HTDS set up a clinic at Lady of Lourdes Hospital in Pasco. On a thyroid ultrasound exam, they found thyroid nodules, but they didn't tell me whether they were cancerous. They didn't biopsy the nodules; they just did an ultrasound on them. The HTDS technicians told me that I was OK, and later on I got a letter from them again saying I was OK. My results were recorded as "no cancer." They told me I had nodules, and in the future I should see a doctor. I learned the thyroid nodules were cancerous in 2003.

I had filed a personal injury claim as a plaintiff in the Hanford downwinder litigation, as a lot of my exposure had occurred when I was a child growing up in Richland. According to the attorneys, based on experts' calculations, I received the highest dose of any of the plaintiffs in the Hanford litigation.

In 2003, my thyroid was removed due to cancer, as was a cancerous lymph node and cancerous parathyroid. The removal of the cancerous parathyroid resulted in hypoparathyroidism and osteoporosis. With hypoparathyroidism, you have to take a lot of calcium, and your body steals the calcium out of your bones. That is what has happened to me.

In 2004 I had prostate cancer. I went through external beam radiation because I didn't want the cancer to spread to other organs. Cancer keeps cropping up due to the damaged DNA in my cells, caused by the radioactivity I have been exposed to. I had forty-three isolated external beam radiation treatments, with two areas of cancer of my prostate treated.

Meanwhile, my COPD continued to worsen, and I went on Social

Security disability. I have oxygen for home use and use a CPAP [continuous positive airway pressure] machine to help me breathe at night.

I have broken the bones in my feet and ribs frequently because of osteoporosis. But because I have neuropathy and can't feel pain, I can still walk—it's probably to my own detriment. In fact, the first time I broke a bone in my foot, in 2006, my new wife, who is an RN, asked me why I was limping, and I said my foot was just a little tender. She said we should go and get it checked out, but I told her not to worry about it. Turns out it was a rotary break, and they had to go in and put a bar and four pins in it. So if I break my feet or break my arms, I still function, even though it definitely doesn't feel comfortable. I don't always feel the break due to the residual neuropathy from spinal meningitis I contracted in my teens.

In 2007, I had a hemorrhoidectomy due to my ongoing colon issues. I also started wearing hearing aids due to damage to my hearing that I believe is from radiation exposure.

In 2008, I had hernia surgery and hydrocele repair. Doctors suspected testicular cancer. It went away with the use of the mineral supplement. A bone-density scan identified advanced osteoporosis. I was put on gout medicine for chronic gout [joint inflammation, a type of arthritis].

After I got over the chondrosarcoma in 2000, I stopped taking the mineral product. I think that was a mistake. I now take it daily. The mineral product works in one of two ways. It either helps my immune system put a sarcophagus around the tumor so it won't expand, or it will knock it out. And in some cases, it does both.

Because of multiple cancers, I have had different organs affected. I have a daily fluid buildup around my lungs for which I take a diuretic. The diuretic has worsened the osteoporosis and caused degeneration of my teeth. The federal nuclear workers' compensation program, called the Energy Employees Occupational Illness Compensation Program (EEOICP), is paying for the diuretic treatment, some of my cancer treatment, and some of the side effects of that treatment. Unfortunately, I have to negotiate with them on each claim, and it takes up to a year to process a claim. This frustrates my doctors.

In 2009, a stent was placed into my left descending heart vein. I believe that my radiation exposure as a child and as a worker at Hanford has affected different organs of the body, including the heart. The cardiologist

I use is also a radiation cardiologist. He works here in Utah. He believes that the heart disease was probably caused by radiation exposure. Twice I have had stents put into my heart because the valves and the veins are not working correctly.

In 2009, an MRI identified five spots of cancer in the spine and left collarbone, deemed inoperable by my surgeon, due in part to the possibility that the surgery would result in paraplegia. I have treated this cancer with the mineral supplement, and the cancerous spots now seem to be under control, as they are not growing.

In 2010, I had cataract surgery in each eye. I believe that the cataracts were due to the effects of radiation exposure. My doctors agree.

In 2013, I again had colon surgery due to blockage, resulting in the removal of approximately ten inches of colon and resection. I also had severe thrush, a side effect of COPD. The thrush has been a continuous problem.

An April 2015 CT scan identified three areas of my small intestine that are probably cancerous. Medication prescribed by my doctor caused multiple bleeding ulcers of the stomach, and I spent four days in the hospital due to massive blood loss. I again used the mineral product and have improved.

In May 2015, cellulitis [a bacterial infection of the skin and layers beneath] and gout resulted in two emergency-room visits and several doctor visits to monitor and treat the condition. Cellulitis on the right arm and elbow had to be surgically treated. I also had cellulitis on my right hand. The gout had progressed to both elbows and hands, both knees, and both feet. This was identified by the rheumatologist as a side effect of the medication I've been taking for COPD.

Like most parents, I feel the need to be around to help out my kids and my seventeen grandkids. I've educated all of them on the importance of watching their health. I've tried to instill a good attitude in them. I have kept them informed on my health issues and progress. The secret is that you have to go on with life.

Look at what they did with the Hanford Thyroid Disease Study. They broke everybody's back. They denied everything they could, and they are still denying it. And, ironically, they use the legal system to beat up the people. But there is one judge they can't hide from . . . that judge we all meet after we leave this life. I feel sorry for them, as they will all answer for everything they say and do, as will we.

I believe in the medical community, but I also believe they are limited by what they have learned in medical school and what they have learned in practice. In the United States, we mask disease rather than treat it. Cancer to me is an immune system breakdown, and you just have to treat it correctly. Helping your immune system to correct damaged DNA/cells is a great way to improve your health.

In your life, things can happen to you. You can be exposed to this or exposed to that. And you can have some health effects of that exposure that challenge what you do. It is how you address those health issues that makes it into either a bad experience or a learning experience. You can find a lot out about the problem and possible approaches to cure it if you have the right attitude and aren't afraid to look.

The Internet is a fantastic resource. I remember when they first brought computers into the Hanford site. They were in the Federal Building downtown. The computer they brought into the Federal Building was a simple computer compared to what we have now. It covered the whole basement of the building. It generated a lot of heat. Today, we are in an information society where we can use our smartphones and go online and find unlimited amounts of things. Computers are so much more compact than those old computers. There are fair and free and honest treatments you can learn about out there that will help you with what is going on.

I went through the standard medical treatments for my first cancer, but if I had it to do again, I would attack the problem from a more knowledgeable, safer perspective. In fact, many oncologists would not go through the radiation and chemo treatments they prescribe for their cancer patients. The pulmonologist I used developed stomach cancer, and went through horrible chemo and radiation treatments. I told him about the mineral supplement. He didn't believe that the supplement would help. He was in his early thirties with two young children. Then I showed him my medical records, and he decided to try the mineral supplement after all. He was back at work for several years. His cancer had vanished. His cancer eventually returned, and he volunteered for a study, trying to help other people out, and went through a bone-marrow transplant. The last call I got from him was from his isolation room about six hours before he

Rance Jones. Photo by author.

Body Snatching and "Project Sunshine"

The AEC's Project Sunshine, which was initiated in 1944 and continued through the 1970s, measured worldwide exposure to radioactive fallout in human populations from nuclear weapons testing. The name referred to researchers' belief that radioactive fallout was as ubiquitous as sunshine.[33]

The project examined human bones and teeth, animals, food, and water for the presence of strontium-90, one of the radionuclides contained in fallout from global testing. Strontium is absorbed into bones like calcium, with more absorbed while bones are still growing in children.

It was very challenging to measure strontium in the living. The preferred approach was to cremate the dead and analyze the resulting ash for radiological content. Human research samples were needed. Under top-secret conditions, hospitals in the United States and abroad supplied dead fetuses and dead babies, as well as the corpses of young people and adults, for study at the University of Chicago within the Enrico Fermi Institute for Nuclear Studies and a satellite research office at Columbia University in New York.[34] The bodies were cremated in the Project Sunshine laboratories, the ashes were measured for radiation content, and the remains were then disposed of.

Over fifteen years, about six thousand corpses were snatched without the knowledge of families. Informed consent was considered "too sensitive to obtain" or "even irrelevant" in the prevailing mentality of the Cold War.[35]

As thermonuclear bomb tests in the Pacific began in 1952, Project Sunshine directors insisted that the research program be accelerated, but they had difficulty obtaining enough human samples. At a biophysics conference convened to discuss the program in light of ongoing thermonuclear bomb testing in the Pacific, AEC Commissioner Dr. Willard Libby told the audience, "Human samples are of prime importance and if anybody knows how to do a good job of body snatching, they will really be serving their country."[36]

The records of Project Sunshine were declassified in 1995 at the request of the President's Advisory Committee on Human Radiation Experiments (ACHRE). The ACHRE concluded that "researchers employed deception in the solicitation of bones of deceased babies from intermediaries with access to human remains."[37]

passed away. Six hours later, his system completely shut down. He had followed the bone-marrow transplant protocol, trying to help others, which required him to stop the mineral supplement.

I am positive. I think it's partly my faith but also knowledge. A lot of the training I got at Hanford and in real estate and investment management has paid off. I would like to get the information out about this mineral product, as I believe it can help individuals with autoimmune diseases, cancers, and other diseases. I believe we can all function at a higher, healthier level if our immune system is strengthened.

Hanford Plutonium Fuels World's First Atomic Test

As soon as Hanford had produced sufficient plutonium for a bomb, that plutonium was used to fuel Trinity, the world's first atomic test, detonated July 16, 1945. The detonation of Trinity's plutonium implosion device showered populations downwind of the New Mexico Test Site with radioactive fallout.[38] Several days after the test, the chief of the medical section of the Manhattan District reported "a very serious hazard" with high dose rates existing throughout a 2,700-square-mile area downwind of the Trinity site.[39]

Prior to the Trinity test, William Laurence, a *New York Times* science writer, had been secretly hired by General Groves. Laurence would serve as the only authorized journalist for the Manhattan Project, tasked with reporting the government's approved version of events. Laurence authored a draft press release that was to be made public following the predawn blast in order to deceive the public about the cause of the explosion and the resultant mushroom cloud that would be clearly visible over the horizon. Laurence's draft press release provided several options for the final wording, depending upon the unpredictable results of the world's first detonation of an atomic bomb. One version, the worst-case scenario, included a listing of the dead. Laurence's final, sanctioned press release reported that "[a] remotely located ammunition magazine containing a considerable number of high explosives and pyrotechnics exploded."[40]

The Trinity test proved the viability of the new plutonium implosion bomb. Less than one month later, on August 6, 1945, Little Boy, fueled by highly enriched uranium-235, decimated Hiroshima, killing or wounding more than 150,000 people. Three days later, Fat Man, fueled by Hanford plutonium-239, destroyed Nagasaki, killing or wounding more than 75,000.

On September 2, 1945, Japan formally signed the instrument of surrender onboard the USS *Missouri* in Tokyo Bay.[41]

"We Do Not Live in a 'City of Pluto'"

From the start-up of the facility in late 1944, Hanford operators kept detailed records on radiation released to the air and to the waters of the

Columbia River, tracking radiation levels off-site in groundwater, on vegetation, and in fish and ducks.[42]

In August 1945, after headlines in the *Richland Villager* first revealed Hanford's mission as a secret atomic bomb production facility of the Manhattan Project, news reports began to question the "health hazards" of Hanford operations.[43] Dr. Herbert Parker was a British American medical physicist who headed up Hanford's Health Physics Division, charged with overseeing Hanford's risk to public health. Parker and Dr. Simon Cantril, a health specialist at the Manhattan Project's Clinton Laboratories in Oak Ridge, Tennessee, who was on temporary assignment at Hanford, responded to these worrisome news stories by issuing a special bulletin to all Hanford employees and their families. The bulletin acknowledged that "recent press and radio releases may cast some doubt in the minds of a few H.E.W. [Hanford Engineer Works][44] personnel concerning the personal safety of themselves and families." Parker and Cantril assured the community that "we do not live in a 'City of Pluto,'"[45] as Richland had been described in some of the reports.

The bulletin assured Hanford families that off-site airborne and waterborne radiation releases were "so insignificant that they could not be measured." With regard to stack releases of gaseous radioactive byproducts of plutonium production, the bulletin assured readers that only a small fraction of the radioactive gases released, measured by Hanford health specialists, returned to ground level. As for the air over Richland, "the quantity [of radioactive gases] is so small that it cannot be measured in the air itself." The bulletin described the levels of radiation detectable in Richland's air as "entirely innocuous" and "approaching the levels of natural activity found in the atmosphere at any location in the country."[46]

In reality, however, radioactive and chemical releases from Hanford frequently far exceeded National Committee on Radiation Protection (NCRP) and International Commission for Radiological Protection (ICRP) tolerance levels.[47] By 1945, Parker had become extremely concerned about high levels of radioactive gas, including I-131, routinely cascading from Hanford's stacks. He devised a means to secretly measure I-131 in the thyroids of cattle in farms surrounding Hanford, a process "which avoided the excitement of public curiosity."[48] Dressed like cowboys, Parker's staff furtively wrangled cattle on area farms "with some difficulty" from army jeeps,[49] roping and throwing the animals, tying them, and then pressing small low-voltage gamma-radiation counters to their necks and throats. Parker's cowboy-attired staff wore earphones connected to the radiation counters so that ranchers and others would not hear the radiation-detecting clicks of the counters and become concerned.[50] The cattle thyroids that were measured registered more than one thousand times the permissible exposure.[51]

Testing revealed that "privately owned herds of cattle ranging to the Columbia River from north of the project area may have accumulated considerable I-131 in their thyroid glands from grazing on contaminated vegetation."[52] Parker's staff collected thyroids from slaughterhouses in Moses Lake, Pasco, Walla Walla, Wenatchee, and Yakima to assess I-131 levels.[53] Covert checks on jackrabbits, waterfowl, and other wildlife began in 1946–1947.[54]

The AEC gave Parker the go-ahead to track radioactive releases and the resulting contamination downwind and downriver of Hanford but not to study the health effects of this contamination. Hanford health physicists "did not, could not, inform the public of higher than permissible levels of contamination as it surged from the plant in random, unpredictable jets."[55]

3

The end of World War II signaled a new era of peace and prosperity for the United States. It was also a time of heightened tension as US leaders grew increasingly concerned that the Soviet Union intended to forcibly spread communism to other nations. At the war's end, the Soviets were intent upon filling the economic and political void in war-ravaged Europe. In March 1946, British prime minister Winston Churchill gave a speech in Fulton, Missouri, referencing an "iron curtain," a "dark shadow of tyranny" cast by an expansionist Soviet Union that had descended across Europe.[1] Three months later, as the first manifestation of the developing arms race between the two nations, the United States began testing atomic bombs, like those dropped on Hiroshima and Nagasaki, in the Pacific Proving Grounds (PPG).[2] Many of these tests were fueled with plutonium produced at Hanford. On July 1, 1946, atomic bombs were dropped on a flotilla of ninety vessels, some with live animals onboard, to study the effects of the blast and the ensuing fallout on animals and naval ships as part of Operation Crossroads.[3]

On August 1, 1946, President Truman signed the Atomic Energy Act (AEA),[4] defining how the United States would control nuclear technology that had been developed in collaboration with the United Kingdom and Canada during the wartime Manhattan Project. Manhattan Project laboratories, personnel, and nuclear weapons production facilities, including Hanford, were transferred to the civilian Atomic Energy Commission (AEC) at midnight, December 31, 1946.[5]

Under a policy that would become known as the Truman Doctrine, President Harry Truman decreed in March 1947 that, if need be, the United States would respond to Soviet postwar aggression with military force. US military commanders pushed for further development, testing, and stockpiling of nuclear weapons. President Truman proclaimed that to ensure the country's nuclear superiority over the Soviets, the United States would develop a nuclear weapon up to a thousand times more powerful than the atomic bombs dropped on Hiroshima and Nagasaki. This new nuclear fusion bomb was referred to as a thermonuclear bomb (also called a hydrogen bomb, H bomb, or Super).[6]

Meanwhile, the Soviet Union continued its acts of aggression. In 1948, the communists overthrew the government of Czechoslovakia, and the Soviets blocked railway, road, and canal access to portions of Berlin under French, British, and US control.[7] Then, on August 29, 1949, the Soviet Union startled the global community with its first test of a fission bomb, referred to by the Americans as "Joe I," referring to Joseph Stalin, dictator of the Soviet Union from 1929 to 1953. The twenty-kiloton bomb was detonated at the Semipalatinsk test site in Kazakhstan. The test occurred well before experts had predicted the Soviets would have an atomic bomb. Americans now began to agonize over the possibility of a Soviet nuclear attack against the United States. Two months later, in October 1949, communist forces overran China.[8]

Richland's Population Explodes

With the end of the war, Hanford faced an uncertain future. Following Truman's speech in March 1947, Hanford employees felt a renewed sense of purpose. On August 14, 1947, triumphant headlines in the *Richland Villager* declared, "Village Faces Boom with Plant Expansion." Hanford families were assured that the facility's continued mission had been secured. Hanford's workforce expanded so quickly in the late 1940s that a new trailer city for construction workers and their families was hurriedly built in North Richland.

The Hanford facility swiftly ramped up operations in response to the perceived Soviet nuclear threat. Additional production reactors were built as engineers and other professionals and their families moved to

Richland. Among them were Lesley and Perry Thompson, who arrived in the fall of 1947.

Lesley's story is the only narrative in this book chronicling health damage to an adult from exposure to Hanford's radiation. Lesley's story provides, with heartbreaking clarity, undeniable proof that older people as well as the young were injured by Hanford's secret off-site radiation emissions.

Lesley is my mother. Her death from malignant melanoma and my father's death from aggressive thyroid cancer devastated me. I believe that my parents' deaths were caused by Hanford's radiation releases. My mother was primarily exposed off-site, while my father was exposed both on the job within Hanford's 100 Area and off-site when not at work.

The motivation to undertake this project, uniting the stories of Hanford's downwind plaintiffs, grew out of my belief that in so doing, my own heart, and possibly our collective hearts as Hanford downwinders, could finally begin to heal.

Plaintiff 6: Lesley Frazier Thompson

Lesley Frazier was born November 25, 1916, in Waitsburg, Washington. She had nearly finished her studies at Washington State University in Pullman when she decided to travel with friends to Hawaii. There, she met Perry Thompson, a navy lieutenant, and they married in Honolulu in 1942, shortly after Japan attacked Pearl Harbor.

In 1947, the couple moved to Richland. Perry was employed by General Electric (GE) as a nuclear engineer in the 100 Area, the location of Hanford's production reactors. Lesley had no real idea what Perry did at his job. When she asked him, he told her he couldn't talk about it.

Richland was a low-crime Mayberry kind of town, populated by scientists, engineers, and other white-collar professionals. Houses in Richland, offered at low rent to families eligible to live within the elite AEC-controlled town,[9] were look-alike structures identified by letters of the alphabet, many with well-tended lawns edged with white picket fences. Lesley and Perry lived in a two-story F house on Stevens Drive, a street lined with newly planted trees.

The AEC offered Richland's residents a wealth of cultural activities in order to mitigate the hardships of life in the secretive community isolated

Lesley and Perry Thompson and their daughter, 1950, Richland, Washington.
Photo courtesy of Thompson family.

within the vast desert expanse of eastern Washington. Lesley enjoyed
Richland's social life, joining the symphony orchestra as a violinist. Perry
tagged along to Richland symphony concerts, although he wasn't a big
fan of classical music and often passed his time doodling on his engineering pad while seated in the back rows of the audience.

Richland seemed like a safe and wonderful place to raise children, and
Lesley and Perry were intent upon starting a family. Their first child, David, born a year after their arrival in Richland, died several days after birth.
The loss of their son devastated the couple, but they tried again two years
later, and this time, the baby, a girl born October 26, 1950, survived.

The family lived in Richland until late 1960, when GE transferred Perry
to its headquarters in San Jose, California, and then to Spain to work on
a reactor being built in the northern part of the country. Several decades
after the family left Richland, Lesley, who had always seemed healthy, developed the first of a series of mysterious health problems. She started to
experience shortness of breath, and her symptoms worsened until she was
diagnosed with severe asthma, an autoimmune disease that required her
to repeatedly use an inhaler to keep her airway open. She had no idea why
she had developed asthma, as no one else in her extended family had ever

been diagnosed with the disease. Then, visual problems cropped up, and by the time she was in her late sixties, she had been diagnosed with both macular degeneration and cataracts. Her vision rapidly deteriorated until she was legally blind.

Lesley heard a news story on the radiation releases from Hanford following the 1986 mass declassification of Hanford records by the DOE. She tried to come to terms with her new understanding that Richland might not have been the utopia it had seemed.

In the late 1980s, Lesley started to feel unremitting fatigue. She underwent a number of tests to determine what was wrong. She was diagnosed with autoimmune thyroiditis (Hashimoto's) and was found to have excessively high blood calcium levels caused by damaged parathyroid glands that were producing too much parathyroid hormone, a disease known as hyperparathyroidism. Then, she began to suffer from chest pain and light-headedness. Tests revealed aortic valve stenosis caused by calcium buildup on the aortic valve in her heart. She underwent risky heart valve replacement surgery, receiving a mechanical heart valve. Blood-thinner medications that she was required to take for the remainder of her life to prevent clotting caused many physical problems, including very easy bruising. Over time, Lesley's skin became paper thin and sloughed off in sheets at the slightest contact, causing profuse, hard-to-control bleeding.

While Lesley was healing from heart surgery, Perry began to experience difficulty swallowing, and it became progressively harder for him to draw deep breaths. Following a battery of medical tests, his doctor suspected cancer and referred him to an oncologist. Perry was diagnosed with aggressive thyroid cancer originating from a nodule on his thyroid. The cancer had metastasized quickly from the thyroid to his esophagus, trachea, lungs, and brain. As Perry's esophagus and trachea became increasingly blocked by metastatic tumors, he also began to lose the ability to swallow. One night, he reached the point where he could barely breathe at all through either his nose or mouth. He was rushed to the emergency room, where a hole was cut through the front of his neck into his trachea and a permanent tracheotomy tube was inserted. He was connected to a breathing machine that forced air in and out of his lungs through a tube. A feeding line was inserted through his flank into his stomach as a means of providing nutrition.

Perry endured both chemotherapy and radiation treatment, struggling to survive the aggressive thyroid cancer. From his hospital bed, he neatly printed messages to his wife and daughter on his engineering pad. He was able to speak a few words if he placed his fingers over the tracheotomy tube, but he refused to do so. His writing became progressively harder to read until, as his condition worsened, it deteriorated into an indecipherable scribble that no one could make out. Neither radiation nor chemotherapy slowed the spread of tumors to his brain, lungs, and other organs. Perry lost his battle against thyroid cancer in March 1996, only months after he was diagnosed.

With his death, Lesley was left without her partner of fifty-four years. She was legally blind and in ill health. She was depressed yet tried to keep her spirits up. Her friends from Richland wrote letters to her in large print they thought she might be able to read, letters that often mentioned friends and neighbors who had succumbed to cancer in the years since she had moved away.

In 1999, Lesley began to suffer from severe back pain and indigestion. Her doctors could not figure out what was happening to her. They decided to run diagnostic tests, and a CT scan showed multiple tumors on her liver. Biopsies revealed that the tumors had originated from a malignant melanoma on her back that had been removed years earlier. Exploratory surgery disclosed that the tumor spread was so extensive that surgery to remove the tumors was not feasible. Her only option was radiation and chemotherapy. She refused to undergo either treatment, citing her husband's valiant but hopeless battle to defeat thyroid cancer. The radiation had burned his skin and caused him to lose his hair, and he had repeatedly suffered through the nausea of chemotherapy. In spite of radiation and chemotherapy, the aggressive cancer had killed him only months after diagnosis. Lesley did not want to go through that. Instead, she chose to die, opting for hospice end-of-life care and medication for pain. She survived for less than three weeks after diagnosis.

Lesley Thompson died in March 1999 of liver failure secondary to aggressive metastatic malignant melanoma.

Hyperparathyroidism and Malignant Melanoma Following Exposure to Ionizing Radiation

Hyperparathyroidism is characterized by excess parathyroid hormone (PTH) in the bloodstream resulting from the overactivity of one or more parathyroid glands, which are three or four tiny glands located in the neck behind the thyroid. The parathyroid glands produce PTH, which maintains calcium levels in the blood and elsewhere in the body. Primary hyperparathyroidism involves the enlargement of one or more parathyroid glands, leading to the overproduction of PTH and resulting in elevated levels of calcium in the blood (hypercalcemia). Hyperparathyroidism can cause a number of significant health problems, including calcification (stenosis) of the aortic valve.[10]

Primary hyperparathyroidism is usually caused by a tumor within the parathyroid glands. An epidemiological study of the incidence of parathyroid tumors in Hiroshima among those exposed to fallout from the atomic bomb showed that the incidence of parathyroid tumors associated with hyperparathyroidism was significantly higher in those exposed than in a nonexposed control group, particularly when they were exposed within 2,000 meters of the hypocenter of the atomic explosion.[11] Studies of people who, between 1939 and 1962 and at less than sixteen years of age, received external-beam radiation treatment to the head and neck for benign conditions also supported an association between radiation exposure and hyperparathyroidism.[12]

The Marshall Islands Nuclear Claims Tribunal Act of 1987, as amended in 1991, adopted regulations establishing a list of medical conditions "irrebuttably presumed to be the result of the Nuclear Testing Program."[13] In 1994, hyperparathyroidism was added to the list of conditions that would enable monetary compensation from the US government for claimants present in the Marshall Islands during the testing program period. The program has run out of funds with many claims still outstanding.

Studies have also suggested a causal connection between malig-

continued

continued from previous page
nant melanoma and exposure to ionizing radiation. At lower levels of exposure, the connection between exposure and malignant melanoma is more uncertain, but evidence shows increased risk approximately proportional to the dose received.[14]

Nuclear worker compensation laws cover malignant melanoma if workers' on-the-job exposure is determined to be significant enough to support causation. Hanford workers applying for compensation for medically verified malignant melanoma under the EEOICPA must show, through a reconstructed occupational exposure dose, that the probability that occupational exposure caused the malignant melanoma is equal to or greater than 50 percent.

Plaintiff 7: Susan Ward

Susan Ward's family, who lived downwind of Hanford during Susan's childhood, did not have a connection to the Hanford facility. Susan was born in Albany, California, in 1946. Her family moved that same year to Walla Walla, where her father began work as an archaeologist for the National Park Service, excavating the ruins of the Whitman Mission and Fort Walla Walla. He also freelanced for the Smithsonian Institution, excavating burial grounds and cremation pits on Sheep Island in the Columbia River near Hanford. Susan and her siblings spent summers with their father, camping near the digs, swimming in the waters of the Columbia, and eating fresh fish caught by members of Native American tribes living in the area. Susan believes her hypothyroidism, basal cell carcinoma (BCC), and several other health issues are related to exposure she received from extensive time outdoors during her childhood downwind and downriver of Hanford.

Here is her story.

My parents met at the University of California, Berkeley, where they both earned combined degrees in archaeology and anthropology. They had lived in Walla Walla before the war, where my sister was born in 1941. In 1942, to help in the war effort, they returned to the Berkeley area, where my dad worked in the Richmond Shipyards. I was born in Albany, California, in 1943.

After the war, in 1946, my family moved back to Walla Walla. There, my dad worked as an archaeologist with the National Park Service. He and his crew excavated the ruins of the historic Whitman Mission, seven miles west of Walla Walla. Also known as Waillatpu, or "place of the rye grass," the mission was established by the Whitmans in 1836 among the Cayuse Indians and was later an important stopping place for immigrants traveling cross-country in wagon trains after completing the difficult crossing of the Blue Mountains. With the immigrants came an epidemic of measles and dysentery, which rapidly spread to the Cayuse, killing many. This may have been the motivation behind the massacre of the founders of the mission, Dr. and Mrs. Marcus Whitman, and twelve others in 1847. The massacre and the destruction of the mission led to many years of war between Native tribes and between Native Americans and white settlers. The mission site and the grave of the massacre victims is now the Whitman National Monument. My dad helped locate the mission buildings and any artifacts still remaining.

He also worked on the excavation of Fort Walla Walla, which during the Whitmans' time was a Hudson's Bay Company fort located near the confluence of the Walla Walla and Columbia Rivers approximately twenty-five miles west of the mission. The fort was founded by the North West Company in 1818. My father and his team located the walls or foundations of six rooms, one of which held a treasure trove of material. They unearthed the largest collection of Hudson's Bay china (whole pieces) in the Northwest. The collection included a washbowl, plates, saucers, pitchers, coffee cups, and other dishware. There were also iron artifacts, large keys, three-tined forks, a whipsaw, and other items. Today, the area once occupied by Fort Walla Walla lies under the waters of the Columbia River due to damming of the river.

From 1949 to 1950, my dad freelanced for the Smithsonian Institution. He excavated ten burials and two cremation pit burials from a village site on Sheep Island in the Columbia River in Benton County, Washington. Sheep Island and Canoe Island were important burial islands for the Imatalamlama and are within the ceded lands of the Confederated Tribes of the Umatilla Indian Reservation. He also excavated at the site of a Native American village at Wallula on the Columbia River. The site was going to be flooded following construction of the McNary Dam, which began in 1947, so he was working to preserve the artifacts there.

My dad got along very well with members of the tribes. He lived with

them for a while and told us that the roasted grasshopper they had prepared for him was really quite delicious. He also told us about sweat lodges
followed by a jump into the icy river. Earlier, in the 1930s, he had lived with
and written an ethnography of the Atsugewi in northern California.

We kids spent time with him in the summer, camping near the digs.
I would often find trading beads. When we found white arrowheads, my
dad confiscated them from us, as they were the arrowheads used by chief-
tains, and he wanted to safekeep them. We spent a lot of time in the river
during the hot summers.

We had a truly amazing childhood, digging in the dirt and discovering
historical artifacts. Little did we know that all along, we were being exposed to radioactive fallout drifting over from Hanford, some eighty-two
miles to the northwest.

I have a memory of myself when I was little, sometime in the winter,
outside our house in Walla Walla. I was wearing my little yellow Mickey
Mouse shirt. I remember getting a really bad sunburn on my face, even
though it was winter. It's a weird memory. This would probably have been
about the time of the Hanford Green Run in the winter of 1949, when Hanford released radioactive iodine and it rained out onto Walla Walla, causing a "hot spot" with high concentrated radiation levels.

We ate fresh vegetables and drank local milk, both of which probably
contained Hanford radiation. Since we lived near a chicken farm, we ate
a lot of "spare parts" from the chickens, like chicken livers. The chickens
would have been eating feed contaminated with Hanford fallout. I know
that livers can retain high levels of toxins, so I think I probably got exposure from eating the chicken livers.

We camped a lot in the summer around my dad's archaeological dig
sites and got fish from the Indians. My dad spent a lot of time with the
tribes, and they even made moccasins for my little sister. We also got Columbia River fish from a neighbor, a local fisherman. That fish included
sturgeon, which are bottom dwellers. I have always wondered whether
the sturgeon ingested radioactive sludge and debris off the bottom of the
river. We drank water from the river when we camped at the archaeological
digs, and we went swimming a lot in the river.

As a kid, I always had cold hands and feet. I was depressed, had bad
teeth and gums and bad fingernails, and was always tired. My mother took

Susan Ward and her siblings at one of their father's archaeological digs. Photo courtesy of Ward family.

me many times to a doctor. I had to take medicine a lot as a child, I remember that.

Around age seventeen, I was the chess master among my friends. I was excellent at strategizing and planning. At age thirty-seven, my IQ tested at 140. A few years later, I developed cognitive issues, described by my doctors as executive function problems. By age fifty, even checkers and chess became difficult for me. I don't know whether these problems were related to my thyroid problems or not.

I had swollen tonsils all the time and finally had a tonsillectomy in 1967.

When I was a young adult, I developed a growth on my neck. In 1981, it was finally diagnosed as a thyroid goiter. I was given thyroid meds to shrink the goiter. After the goiter shrank, I went off the meds, which caused me to become nervous, depressed, spacey, and unable to concentrate. After some time off the thyroid meds, the goiter returned, which made my depression worsen, as I was very worried about what was happening with my health.

Around 1990, an endocrinologist performed a series of diagnostic procedures to try to figure out what was going on. He told me that my thyroid was misshapen. The endocrinologist referred me to nuclear medicine for a thyroid scan. I had to go off thyroid meds for a week for the scan, which again made me feel really bad.

The scan and a fine-needle aspiration revealed thyroid nodules. The nodules combined with the goiter were distressing to me. I continue to have nodules on my thyroid. I am very concerned about them. I am monitored about once a year, and I have had some of them biopsied and removed. I'm hypothyroid now and again on synthetic thyroid meds. I worry about thyroid cancer in the future.

Hypothyroidism has caused me to have low energy and chronic depression. My nails are horrible, my concentration is not good, and I have poor memory. My skin, eyes, and gums are extra dry. Dry eyes are painful, and dry gums can lead to serious dental problems. It's bad for the bones in the jaw, and I've had lots of cavities. I have muscle spasms, and I don't know whether they are related to thyroid disease or not. I also have pretty severe osteoarthritis.

I always had very irregular periods. I had scant menses, alternating with sudden prolific menses. I sometimes fainted, possibly due to anemia from blood loss. I ended up having a hysterectomy at age twenty-seven.

I have gluten sensitivity and poor digestion. I sometimes have trouble with choking, even when not eating or drinking. I have had gastroesophageal reflux disease (GERD) since I was young. The GERD has not been helped by thyroid medication.

I also have basal cell carcinoma, which was diagnosed in 1993, with spiderlike growths on one side of my face and on my ear. These growths started showing up when I was about forty years old. I went to a doctor, who said the carcinoma was related to either radiation or strychnine exposure and that it couldn't have been caused by sun exposure. I haven't been exposed to strychnine.

He said that mine was a "textbook" case following radiation exposure. He said that the exposure does something to your cells. Twenty to thirty years after exposure, the growths start to show up. This cancer was highly distressing to me. Parts of my face would start bleeding. I had the basal cells removed, and then they would grow back in the same area. It was very painful. This went on for about five years until it finally stopped. By that time, I had gone through thirty surgeries to try to restore the end of my

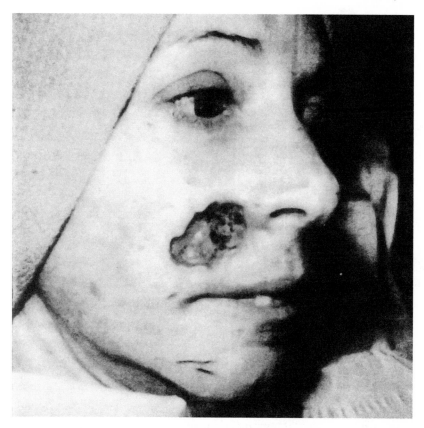

Susan Ward: facial disfigurement from basal cell carcinoma. Photo courtesy of Ward family.

nose and for other facial reconstruction. My cheeks were also deformed by the basal cell growths. Chunks had been taken out of my cheeks to remove the growths. The cancer treatment caused chronic pain. I started taking antidepressants because of the deformity.

I had holes in my face and part of my nose cut off. It was disfiguring and painful. I had to take time off from work. Because of the disfigurement of my nose and face, I was only working nights while in bandages so that I didn't have to see people or have people see me as much.

I worry I may have passed something on to my children due to my Hanford radiation exposure. Both of my sons have type 1 diabetes and migraine headaches, and my oldest son has vision loss from retinopathy. My daughter has thyroid disease, Graves' orbitopathy, and bone cysts.

Susan Ward. Photo courtesy of Ward family.

I have five siblings, several of whom have health problems. My oldest sister, born in Walla Walla in 1941, has thyroid issues and was recently diagnosed with frontal temporal dementia. Her first child was stillborn. My sister born in Walla Walla in 1949 has all kinds of health issues. One of my brothers was born in 1953 with Down syndrome.

I had a miscarriage in 1960. I know a lot of women in the Hanford downwind area who had miscarriages around that time.

My father had breast cancer, kidney disease, and osteoarthritis.

All the health problems I have suffered have had a major impact on my life. In 2011, I had to quit working as a clinical nurse specialist because of severe osteoarthritis and disabling muscle spasms.

When my doctor told me I had been exposed to radiation, I started researching skin cancers and radiation and found a connection, especially to BCC. The thyroid problems caused their own set of issues. I have been dealing with these problems for so long that it is hard to say what "normal" is.

One of the things that really bothers me about what has happened to the Hanford downwinders is the lack of accountability by our government. I have always thought an official apology would be helpful and appropriate. I would not have known about Hanford at all if it hadn't been in the newspapers in California, where I have lived a long time. They should have let everyone know what happened long before they did.

Native Americans and Farm Workers

While camping near the archaeological digs, Susan and her siblings frequently interacted with Native Americans living near Walla Walla and the Hanford Reach of the Columbia.

Native American tribes and nations, including the Colville, Coeur d'Alene, Kootenai, Spokane, Warm Springs, Klickitat, Yakama, Nez Percé, Umatilla, Kalispel, Cayuse, and Walla Walla,[15] resided downwind or downriver of Hanford during decades of radiation releases from the facility. In 2004, the Kalispel tribe conducted a health survey of health issues among tribal members that may have resulted from Hanford radiation exposure.[16] The survey focused on thyroid issues, reproductive cancers, and other health problems.

The Hanford Environmental Dose Reconstruction (HEDR) project revealed that a number of Native American communities may have received higher exposures than non-Native people in the downwind area due to their traditional hunting, gathering, and fishing practices.[17] Hanford-area tribal families often consumed high quantities of whole fish caught in the radiation-polluted Columbia River.

A class action suit filed in 1997 in the US District Court, Western District of Washington, on behalf of individual Native Americans downwind of Hanford alleged that Native Americans were deliberately exposed to radiation from Hanford as part of secret government radiological tests during the Cold War. Named defendants included the federal government, several former Hanford contractors, Battelle Pacific Northwest Laboratory, and the University of Washington, which had conducted research for the AEC at Hanford. Pleadings in the case asserted that Native Americans were at substantial risk from Hanford radiation releases due to lifestyle, dietary and religious practices, and cultural habits that involved spending a great deal of time outdoors.[18]

In 1989, the Environmental Protection Agency (EPA) began a study of Native American exposures downwind of Hanford. The study, released in 2002, concluded that Columbia River tribes in the Hanford region faced a high risk of cancer and other disease due to toxic pollutants in fish: "For Native Americans eating the most salmon, steelhead and rainbow trout, the risk of developing cancer ranged from 7 cases in every 10,000 people to 2 in 1000, based on tests of the fish caught at different locations in the Columbia Basin."[19] Individuals consuming long-lived resident fish varieties such as sturgeon risked a two-in-one-hundred chance of cancer compared to the general population at some locations in the basin.[20]

The study suggested that Native Americans may have had more exposure to Hanford radiation than previously believed and that the HEDR estimates for Native Americans may have been too low. They may have eaten more fish than other people living downwind and downriver of Hanford and would have prepared the fish in a way that exposed them to potentially cancer-causing radiation. The HEDR did not take into account radioactive strontium, which concentrates in fish bones. Native families often boiled fish bones in stewpots after the fish were smoked, and boiling released the radioactive strontium.[21]

In August 2002, Indian tribes and the EPA jointly conducted the Columbia River Basin Fish Contaminant Survey. The study found that nonradioactive contaminants in fish, including polychlorinated biphenyls (PCBs), are found in the highest levels in the area of the Columbia River that runs through the Hanford Reach, the portion of the river that flowed through the Hanford site. The study found that tribal children who ate fish from the Hanford Reach had a one hundred times greater risk of immune diseases and central nervous system disorders than non-Indian children. The risk to tribal members of developing cancers from eating fish from the Hanford Reach was estimated to be as high as one in fifty. While the DOE blamed Canadian metal mines upriver on the Columbia for the high levels of nonradioactive contaminants in the river, the tribes believed the source to be Hanford.[22]

Farm workers in the Hanford region also spent extensive time outdoors and may have been exposed to significant levels of radiation from Hanford, much like the Native American communities. Farm labor deficits at the time of the Hanford start-up in late 1944 resulted from the combined effects of wartime transportation restrictions, the draft, and workers moving to higher-paying jobs in the defense industry. To address the labor shortage, special populations were tapped, including Mexican nationals brought into the United States under the "bracero" program; German prisoners of war; Japanese internees; state and local prisoners; American soldiers on leave; Canadian indigenous peoples; youths between the ages of fourteen and eighteen in the Victory Farm Volunteer Program; and women, children, and other civilians under the Women's Land Army Program. All of these programs contributed farm workers within the Hanford downwind areas. These temporary workers spent extensive time outside during planting and harvesting and may have received substantial exposure to radiation released at the time from the Hanford facility.[23]

North Richland/Camp Hanford

Work on the North Richland Construction Camp, four miles north of Richland, which would eventually house twenty-five thousand construction workers and their families, was completed in 1948 as part of Hanford's massive postwar expansion. Once the camp was ready for occupancy, amid raging sandstorms, sixteen thousand workers accompanied by ten thousand family members moved into barracks, trailers, and prefabricated homes. The camp featured a mess hall, school, post office, fire station, drug and grocery stores, and police patrol headquarters. In 1950, the "biggest convoy in Washington State's history"[24] arrived in Hanford from Fort Lewis, southwest of Tacoma, to protect the reactors and chemical separations plants from potential Soviet air attack. Antiaircraft batteries were positioned throughout the Hanford site, with headquarters in North Richland. In 1951, the military compound was designated Camp Hanford.[25]

In the late 1950s, the antiaircraft batteries were replaced by four Nike missile installations on the Hanford site and three more in the nearby area. By 1960, Nike missiles had become outdated. Camp Hanford was officially closed on March 31, 1961. The buildings and campsite were dismantled between 1962 and 1965, and portions of the area were returned to farming use. The site is now the location of the DOE's Pacific Northwest National Laboratory, Washington State University, and other institutions.

Plaintiff 8: Bonnie Rae

Bonnie Rae was born in Walla Walla, Washington, in 1947. When Bonnie was four years old, her father went to Hanford as a "Phase II" construction worker. Bonnie's family initially lived in the trailer camps of North Richland, moving later to Pasco and then to Kennewick. Bonnie's mother, who worked at Lady of Lourdes Hospital in Pasco, spoke often to Bonnie about the families that came to the hospital in the 1950s seeking treatment for children who were acutely ill with leukemia. Bonnie also remembers friends with siblings who developed leukemia. Bonnie blames her own health issues on her childhood exposures in the Tri-Cities.

This is her story.

I was born in Walla Walla in 1947. We lived right next door to a dairy farm, so all our milk was fresh and unpasteurized. My mother had a huge vegetable garden, and my dad was a big hunter and fisherman, so our food in those days came from those sources.

We left Walla Walla in 1951, when I was four. I spent a year in North Richland before I started school. We were part of what they called the "Phase II Hanford workers." Phase I was, of course, building the atomic bomb during the wartime Manhattan Project. Hanford just about closed down after its plutonium was used in the Nagasaki bomb on August 9, 1945. They had started getting rid of the soldiers who guarded the site and everything. Then, the Soviet Union fired off its first atomic test at the end of 1949, and suddenly, Hanford was booming again and was even bigger than before.

My dad was a pipe fitter. We lived in the huge trailer camp in North Richland. On June 1, 1955, the government decided to kick out twenty-five thousand people who lived in the trailer camp. They gave us all about a month to move. We moved to Pasco for a year, and then my folks built a house in Kennewick. My dad continued to work in construction.

My dad was bothered about the radiation. He would say, "All that radiation couldn't be doing anyone any good." He sometimes spoke about an accident out there when the "grass turned black." This must have been around 1952 or 1953. Thirty or forty years later, I went to a meeting of HEAL in Spokane and heard an old-timer talk about that same accident when the "grass turned black." Yes, it's anecdotal, but something happened. Yet they act like nothing happened.

Bonnie Rae as a child. Photo courtesy of Rae family.

When I was growing up, my mom worked at Lady of Lourdes Hospital in Pasco, and childhood leukemia was common. I know all kinds of kids who had little brothers and sisters who developed leukemia. It was always being talked about. You can see all the graves of children from that time who died from childhood leukemia if you visit the historic Richland Cemetery.

Nodules appeared on my thyroid when I was pregnant at eighteen with my first child. The doctor was not really worried about the nodules. In 1968, just weeks after I had my second daughter, I was brought in for a subtotal thyroidectomy, as the doctor this time was very concerned. Back then, when they found nodules, they routinely took out the thyroid lobe containing the nodules. They didn't do a biopsy or ultrasound or have access to most of the technology they have today. The scar across my neck was scarlet red. The doctor said to rub the surgery scar with hand lotion, but it hurt to do that.

Other than the nodules and the surgery and ongoing problems with my thyroid, I'm pretty healthy. I seemed to be OK for years and years, but now my thyroid-stimulating hormone (TSH) is low, which means I'm hyperthyroid.

I'm convinced my thyroid problems began with the Green Run. The Green Run, December 2–3, 1949, was a secret experiment that released thousands of curies of I-131. They said the purpose of the Green Run was to test monitoring equipment the air force was developing to track Soviet nuclear tests. The injustice of it is what bothers me. It was an intentional release. What happened to me may have happened to every other toddler at the time in Walla Walla. That has bothered me ever since I learned about it in 1986. The Green Run was about a thousand times worse than Three Mile Island, and yet there, everyone was evacuated, especially the children.

My husband, who died recently, lived in Spokane as a child. I know Hanford's radiation came into Spokane. I don't know whether his death was connected to exposure to Hanford fallout that reached Spokane.

The federal government ran the HEDR, and the dose they calculated for me was low. That was ridiculous. I lived next door to a dairy farm, drank fresh milk, not pasteurized. That should have made my dose high. The I-131 goes into the milk, and when that milk is fresh rather than pasteurized, the I-131 levels are still high when people drink it. Someone should have been held accountable for that dose reconstruction project.

My dad probably didn't have any health problems relating to Hanford. He died of a heart attack. My mother was having thyroid problems while we were in North Richland. She went to Virginia Mason Clinic in Seattle in 1952 and supposedly had a nervous breakdown. She also got as skinny as

Bonnie Rae. Photo by author.

hell. I remember being outside the clinic trying to visit her. They wouldn't let children in, and I remember seeing her in the window on the second or third floor. I don't remember how long she was in there—something happened with her health.

She died in 1989 of an extremely rare cancer that I feel was related to Hanford. She had worked at Hanford as a clerk typist. It was a malignant melanoma that started internally. They called it "atypical site melanoma," which is a rare form of melanoma. It metastasized all over her body very fast. They took out everything they could, but they couldn't stop it.

What I really don't understand is why the litigation took twenty-four years. They compensate Hanford workers for cancers and other diseases, but not us. It didn't take that many years for the people in St. George, Utah, to get compensated for their exposures to Nevada Test Site fallout!

Some government official ought to send me a letter to tell me yes, they did it deliberately, the Green Run. And they need to apologize.

I'd like people to be aware that the government of the United States is not always on the citizens' side. The Green Run was a deliberate act of aggression, and the people of eastern Washington were harmed deliberately. People in eastern Washington were guinea pigs for the government.

Up until the day Emperor Hirohito signed the peace agreement on the battleship *Missouri*, whatever Hanford did is A-OK by me. I'm not against the bomb per se since that saved millions of lives, but after that day, I don't understand why we continued to produce atomic weapons. Russia was our ally, and we couldn't have won the war without Russia. And they took the brunt of the casualties. And then to make more nuclear weapons for possible use against Russia, weapons that are still in existence that could blow up the earth a hundred times over—that is foolish, wrong, and we ought to be talking more about it.

The Green Run Experiment

After the Soviets tested their first atomic bomb in August 1949, the United States raced to figure out a means to determine how many more atomic bombs the Soviets were producing. One effort to verify the Soviets' rate of production was the Hanford Green Run, part of a classified program in the late 1940s and 1950s code-named Operation Blue Nose that was run jointly by the AEC and the US Air Force.[26] The goal of the program was to try to locate plutonium production plants within the Soviet Union and to measure Soviet plutonium production levels through the analysis of radioactive fission-product gases emitted during the reprocessing of reactor fuel.[27]

The Green Run was a secret experiment conducted on December 2–3, 1949, involving the airborne discharge from the Hanford T[28] of radioactive fission products from processing one ton of uranium fuel slugs irradiated in Hanford's reactors and then cooled for only sixteen days rather than the usual much longer period of up to 101 days, which would have allowed radioactivity to decay. The radioactive gases released from Hanford during the Green Run were still "green," highly radioactive,[29] and enormously dangerous to communities downwind.

The Green Run was conducted by GE's Nucleonics Department at Hanford in spite of Hanford officials' concern that poor weather at the time would unnecessarily expose downwind populations. Hanford scientists calculated the expected Green Run releases at about 4,000 curies of I-131 and 7,900 curies of xenon-133, although the measured values were higher than the estimates by a factor of two to three.[30] Vegetation samples in Kennewick revealed I-131 levels over 1,000 times the (then) acceptable levels.[31]

According to one of Hanford's leading radiation control managers at the time of the Green Run, the AEC assumed that since the first Soviet atomic test, the Soviets were rushing to produce atomic bombs using "green" fuel about twenty days out of the reactor—and that this "green fuel" should be radioactive enough to be easily detected by the United States.[32]

US monitoring stations located just outside the Soviet Union could measure the airborne radioactive gases being released. Air force officials wanted to know what the gas levels meant in terms of the amount of uranium being processed by the Soviets, so they decided to operate Hanford's T plant "Soviet style," shortening the cooling periods and allowing higher levels of radioactive gases to be released from the Hanford stacks. They would then measure off-site radiation and apply a formula that allowed scientists to convert the amount of radioactive gas detected by monitoring stations to estimate the Soviet bomb production rate. The Soviets processed green uranium, so Hanford would also process green uranium for this test.[33]

The Green Run released two to three times more I-131 and other radioactive gases than the planners of the experiment had anticipated.[34] The public was not warned, and Hanford managers did not conduct follow-up studies of the health of people exposed.

Following the Green Run, the levels of radioactive fallout deposited in the Hanford area were higher than projected because of a moderate temperature inversion and calm winds at the time. The thyroid glands of animals collected from the area around Hanford at the time of the experiment contained up to eighty times the "tolerance levels" defined by Hanford scientists for continuous exposure. According to Hanford records, "The test was conducted despite less-than-optimal weather conditions, which ... may have exposed greater-than-expected numbers of the population to the radioactive cloud. Prevailing wind patterns prior to the test had been inopportune, and wind shifts during the test caused the emission of gases close to the ground, including directional shifts over populated areas in southeast Washington."[35]

On December 2, 1949, the weather in Walla Walla, located less than eighty-four miles southeast of Hanford, was calm, changing to snow and ice pellets on December 3.[36] The combination of calm air on December 2 followed by snow and ice the following day caused a radioactive hot spot to develop in Walla Walla, subjecting people in the area to substantially higher exposures than those received in surrounding areas. Spokane is also believed to have been a hot spot for the Green Run and other Hanford releases due to winds that combined plumes of radiation over the area.[37]

At the time of the Green Run, Connie Nelson was only one year old, growing up in the small town of Garfield, south of Spokane. After she got married, she and her husband tried for eight years to become pregnant. At a consultation, she showed her obstetrician a lump on her neck that was found to be malignant. June Stark Casey was a young student at Whitman College in Walla Walla at the time of the Green Run in December 1949. She began to feel unwell in early 1950, and doctors found that she had become extremely hypothyroid, a health issue she had never faced before. She later faced reproductive problems that she feels were likely related to her exposure during the Green Run.

Here are their stories.

Plaintiff 9: Connie Nelson

I was born in Spokane in 1948. My family lived in a small town south of Spokane called Garfield.

In Garfield, my family had a little farm. We had a backyard cow and drank our milk directly from the cow. It wasn't boiled or anything. When I was little, I used to watch my dad milk the cows while I sat in my little car seat next to him. We kids also crawled around in the dirt and pulled the carrots out of the ground to eat. We had a vegetable garden, and we ate the vegetables straight out of the garden.

I stayed in Garfield until I graduated from high school in 1967. Then, I went to Eastern Washington University in Cheney, graduating in 1971.

I worked in Spokane for a year, then went back to school in Ellensburg and got a career and technical education degree. I started teaching in September 1973 in a little town called Anatone, the most southeastern town in the state of Washington, above Clarkston in the Blue Mountains.

I got married in 1976. After almost eight years of unsuccessfully trying

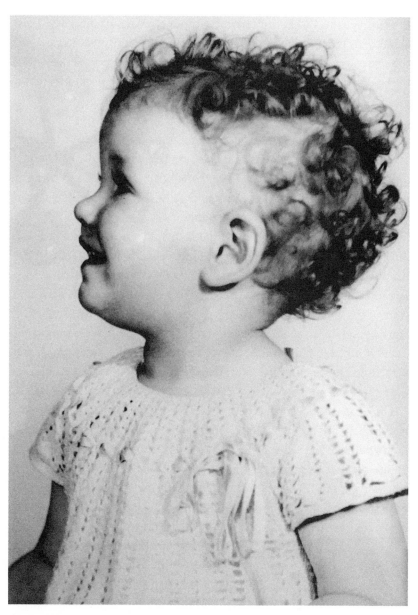

Connie Nelson as a young child. Photo courtesy of Nelson family.

to get pregnant, I went in to have my tubes checked for blockage. I also showed my obstetrician a lump in a lymph node on my neck. It was about the size of a pea. He said, "I don't like that!" So after dye was injected and they didn't find any blockage in my fallopian tubes, they sent me right to an endocrinologist.

The endocrinologist thought at first that the lump on my neck was cat scratch fever. He sent me to a doctor, and they did a biopsy of the lump and found that it was malignant. This was February 1983. I had surgery quite soon thereafter. A biopsy determined that the tumor had metasta-sized from my thyroid. They removed the entire thyroid but managed to preserve the parathyroids, thank goodness.

They immediately put me on synthetic thyroid hormone, and then they did a lot of blood tests. I have deep rolling veins, so it's very hard to give blood. I will have to take synthetic thyroid for the rest of my life.

I had to go back for several more radiation treatments. I was trying to work at the same time. I had already done a couple of rounds of radiation, and when I went in again, they asked me if I could possibly be pregnant. I didn't think so, but they said, "Let's just do a test before we treat you again with radiation"—and, what a surprise, I was pregnant! They discontinued the treatments.

I had an amniocentesis, and everything was normal. I delivered a healthy baby girl in August 1984. Unfortunately, because I needed another radiation treatment after her birth, I breastfed for only about six weeks. Before my next radiation treatment, I went to the pediatrician and asked whether I could breastfeed after treatment. She thought I should stop breastfeeding for about a week, but then I could continue. I then went to my endocrinologist and asked the same question. He advised that I stop for two weeks. When I went in to have the radiation treatment, I brought my baby, who was about six to eight weeks old. The technician asked if I was breastfeeding. She said she hoped I would no longer breastfeed. It also concerned her that I had the baby with me due to possible radiation exposure from my treatment.

Before I knew I was pregnant, I had gone to the hospital, where I drank a vial of strange liquid. I now know it was probably radioactive iodine. The liquid arrived with a guy with a full white suit with a lead container. It was so scary. My husband set it up so the windshield and radio came on as a radiation gag when I got back in the car. . . . He was just trying to use humor to make things a little easier to handle.

After the birth and all the radiation treatments, I had to go off all my thyroid meds for at least a week so they could check my thyroid again. I had to wean myself off the meds. I got headaches and felt pressure in my head. I was exhausted. And I was trying to care for a baby!

After the surgery in 1983, I started to wonder why I got thyroid cancer. I was living in Shelton, Washington, at the time, way over in western Washington.

Around 1986, my aunt in Rosalia, a little south of Spokane, told me about Hanford. There wasn't anything in the news about Hanford in Shelton, where I lived. My aunt said there was this article in a Spokane newspaper that gave Tom Foulds's number. So I called him up. When the settlements finally came about in 2015, it was my friend in Boise who called me and told me about it, as there wasn't anything about the settlements in the local papers either.

My mom was still alive at the time I filed as a plaintiff, and she was able to help me fill out the interrogatories for the litigation, including all the dietary information about what I ate as a kid, how many servings of milk I drank, and whether the milk was fresh or pasteurized from a dairy.

I was the only one of the twenty-one in my graduating class at Garfield High School who had thyroid cancer. A couple of them had goiters. Most of them still didn't know about Hanford radiation releases. Garfield is over by the Idaho border, so news about Hanford isn't in the newspaper there.

My dad didn't have thyroid issues. He died early, at age sixty-one, of heart disease. I don't know if it was related to Hanford. My mom took thyroid meds her whole life. Between 1942 and when I was born in 1948, she had eight miscarriages. Maybe some of them were related to Hanford.

The infertility issues were so hard to face. To be an only child, then to have only one child after eight years of trying was so hard. My husband has two siblings, and we wanted more than one child, but we were only able to have one.

Just knowing that I have had thyroid cancer and worrying about it returning is also very difficult.

I was angry at first, but over the years, I guess I've gotten used to it. I go in and get checked regularly. The weight issue is tough as well. It's so

Connie Nelson. Photo by author.

easy to put on weight with thyroid problems. Your whole metabolism is determined by the one little thyroid gland. I have had knee issues relating to the weight. It's kind of like a domino effect.

The story of the downwinders and the damage done to them needs to be known. The damage was done to normal citizens. How dare they hide this under the carpet? Why did it take twenty-four years of litigation to get this far? We have a lifelong disease. The government has to be made aware that you can't just do that and expect people to just deal with it. That is not the way people should be treated.

We are like a sacrificed group. People who haven't been through this can't really relate to my story.

Plaintiff 10: June Stark Casey

I was born in 1930 in Portland, Oregon. My family moved to The Dalles, Oregon, where I lived until I went off to attend college as a declared music major.

I went to Seattle University my first year. My mother learned that the

June Stark Casey as a young adult. Photo courtesy of Casey family.

best music school on the West Coast was Whitman College in Walla Walla, Washington. I decided to transfer to Whitman and moved to Walla Walla to begin my studies there in September of 1949, when I was nineteen. I stayed in Walla Walla two years. I then transferred to the University of Washington, from which I graduated.

Hanford's Green Run experiment took place December 2–3, 1949.

Hanford released thousands of curies of I-131 and 16,000 to 24,000 curies of xenon-133. A radiation hot spot was created in Walla Walla when calm winds on December 2 caused the radiation plume to remain over the city, then snow and ice pellets the next day caused the radiation to be deposited on the ground and everything else.

I immediately noticed health issues in December of that year. I went home to The Dalles for Christmas vacation. I had begun to feel very fatigued and not well in general. I thought maybe it was because I was studying so hard. My mother suspected something was wrong with me, and she took me to the doctor. I was given a thyroid blood test, and they found very high TSH levels, indicators of hypothyroidism. I didn't have hypothyroidism or any kind of thyroid problems before I moved to Walla Walla!

I was immediately put on synthetic thyroid hormone, but I don't recall which brand. I started to lose my hair. My hair never grew back, and now I wear a wig.

I married, then suffered both a miscarriage and a stillbirth. I think this was probably due to Hanford radiation, as it is well known that radiation exposure can cause reproductive problems. Amazingly, I had a son before all these reproductive issues started up. He was a miracle.

In addition to the hypothyroidism, I have had other health problems that I feel were caused by my exposure during the Green Run. I have breast cancer that was first diagnosed in 1997. The cancer was surgically excised. The cancer recurred in 2015, at which time I had surgery in both breasts and lymph nodes, followed by thirty radiation treatments.

I have also had skin cancer, chronic degenerative spine disease, and thyroid nodules. Esophageal problems make it difficult to keep food down and have required two endoscopies. Unremitting pain in my spine makes it feel like someone is torturing me with an electric saw. My shoulders feel as if someone is holding them over hot coals in a barbeque pit. I have medication now that is helping with the spinal pain, but I am afraid if I stop it, the pain will return.

I live in Oakland, California. I learned about Hanford from an article in the Oakland Tribune that I read on Mother's Day, 1986. There was an article about the Green Run and Hanford. Reading it, I realized the connection to my health problems. I was in tears. My parents went to their graves not

June Stark Casey. Photo by author.

knowing about Hanford. I do remember that my mother said everything about Hanford was hush-hush, but she hadn't a clue that I had been exposed while in Walla Walla.

Over the years, I have done a lot of charity work and have served on many boards. Once I realized that many, if not all, of my health issues were caused by my exposure to radiation released during the Green Run, I devoted myself to antinuclear activism.

I coordinated two conferences in Richland and Seattle, Washington, in 2000 and 2005 for the Japan Council against A and H bombs prior to the Japanese delegation's attendance at the Nuclear Non-Proliferation Treaty Review Conference in New York City. I put together a huge booklet of articles about Hanford for them to take with them.

My story has been included in six documentaries, including *Deadly Deception*, an Oscar-winning short documentary. I have spoken at US and international conferences, at DOE hearings, at GE shareholder meetings, and on radio. The most meaningful presentations for me were in Hiroshima and Nagasaki at the World Conference against Atomic and Hydrogen Bombs and in Cairo with the Afro-Asian Peoples' Solidarity Organization.

Once I learned of my exposure during the Green Run, I filed as a plaintiff in the Hanford litigation, but my case was later dismissed. Apparently, my I-131 exposure dose calculated by the HEDR wasn't high enough for the lawyers to prove that my thyroid issues were caused by Hanford exposure. How do they really know what I was exposed to there in Walla Walla, a known hot spot of Hanford radiation? I feel that had I not moved to Walla Walla during the Green Run, I wouldn't have the health issues I now have. The litigation didn't even address all the other health issues so many Hanford downwinders have—the reproductive problems, birth defects, autoimmune disease, other cancers, and other serious health issues.

I want people to know what happened to all of us. We can't rely on the DOE to tell the truth. Some people think that all of this happened so long ago that it doesn't matter anymore. It does.

Project Nutmeg: The Search for a New Continental Test Site

In 1948, as part of Project Nutmeg, the AEC undertook a search for a new continental atomic test site. The original continental test site where Trinity had been detonated on July 16, 1945, was located in a remote corner of the Alamogordo Bombing Range within the Jornada del Muerto desert, about 210 miles south of Los Alamos, New Mexico. Nuclear testing at the PPG had presented monumental logistical challenges related to the transport, housing, and supply of test personnel. The humid climate of the Pacific had also created problems for both electronic and photographic equipment. The AEC sought a site where security and accessibility would not be jeopardized by enemy action.[38] These concerns intensified on June 25, 1950, when North Korean communist troops, supported by the Soviets, crossed the Thirty-Eighth Parallel and invaded South Korea, triggering the Korean War. President Truman committed troops to a UN military response. The Korean War was the first major postwar conflict engaged in by the United States with the goal of containing Russian expansion. The search for a new test site was eventually narrowed to a location within the Nevada-Toponah Bombing and Gunnery Range, controlled by the air force. The site, originally known as the Nevada Proving Grounds (NPG) and renamed the Nevada Test Site (NTS) in 1955, is located in Nye County, some sixty-five miles northwest of Las Vegas, Nevada. The location was chosen in spite of serious concern about the possibility of radioactive fallout from tests, which was projected to travel up to 125 miles downwind.[39]

On December 18, 1950, President Truman authorized the new test site for the detonation of atomic fission weapons, weapon prototypes, or experimental devices of up to fifty kilotons.[40] By this time, thousands of civilians downwind of the Trinity Test Site, Hanford, the PPG, and several other Manhattan Project atomic production sites had been exposed to fallout from the US nuclear weapons program. None of these civilians were warned that the US government had placed them in harm's way. The United States failed to monitor the radiation exposure of individuals downwind and did not study the immediate or long-term health damage resulting from exposure. With the initiation of aboveground testing at the new continental test site, the number of exposure victims would vastly increase.[41]

4

Well before the start of atmospheric (aboveground) atomic testing in January 1951 at the NPG, the AEC was aware that the tests would expose downwind populations to radioactive fallout.[1] Preliminary results from studies of survivors of the atomic bombings of Hiroshima and Nagasaki had found a substantial increase in the incidence of leukemia in the exposed population when compared to a control group, although this information had not been made public by the AEC.[2] The scientific community at the time recognized that exposure to radiation could cause leukemia, other cancers, genetic defects, and other health issues.[3]

Risks from fallout included external exposure to gamma and beta radiation; internal exposure through ingestion or absorption of alpha-, beta-, and gamma-emitting radionuclides; and inhalation of fine particles of descending and resuspended radioactive debris from atomic detonations, especially those containing high-energy alpha emitters, including plutonium.[4] Many of the tests at the NPG were detonated just above the surface of the test site on towers or balloons, sucking up dirt and debris that became radioactive and resulting in significant fallout, particularly in areas closer to the test site.

Some of the radioisotopes in fallout were known to concentrate in sensitive organs and tissues or could persist in the environment for long periods of time. Inhalation or ingestion of tiny amounts of plutonium-239, strontium-90, and similar radioactive substances could lead to leukemia and other cancers, while I-131 and other radioactive isotopes of iodine in fallout could concentrate in the thyroid gland and other organs.[5]

During atmospheric testing at the continental test site, from 1951 until 1963, when testing moved underground, the AEC made no effort to measure internal exposures from fallout to people downwind. At the time of the first air-dropped tests, no one knew how to monitor the radiation released.[6] The first tests were scheduled so quickly that there was no time to establish a network of monitoring stations. Instead, a handful of radiation monitoring personnel made an effort to track the path of fallout clouds, ineffectually attempting to cover a vast amount of territory.

During Buster Jangle, the second series of tests that began in October 1951, chunks of sticky paper were placed on fence posts to collect radioactive fallout particles. Off-site monitoring personnel went where meteorologists predicted the fallout clouds would travel, carrying air samplers made out of Electrolux vacuum cleaners to capture radiation in the vacuum filters.[7] However, variable wind patterns often caused fallout clouds to travel in unanticipated directions, far from the location of the monitoring personnel.

From 1951 through 1962, monitoring focused primarily on the first, acute phase of fallout hazards from tests, measuring external gamma exposure for a few hours immediately following detonations while ignoring long-term exposure. Until the Plumbbob shot in 1957, chronic, long-term exposures were largely ignored. Off-site monitors were sometimes told to keep beta detection shields on monitoring devices closed, preventing the measurement of beta-emitting particles.[8] Dose estimation for downwind populations was "at best an educated guess."[9]

Thermonuclear tests began in the PPG in 1952. These immense detonations, far more powerful than the atomic tests conducted at the NPG, exposed indigenous people on downwind atolls, American military personnel and others witnessing the tests, and people accidentally caught in the path of fallout to extremely high levels of radiation.[10] The Japanese fishing boat *Daigo Fukuryū Maru* (Lucky Dragon No. 5) was caught in fallout from the Castle Bravo thermonuclear test in 1954, killing its radioman. His last words were "I pray that I am the last victim of an atomic or hydrogen bomb."[11]

Fallout Moves East

Atomic tests at the NPG spread airborne fallout throughout the United States. Shortly after atmospheric testing began in 1951, the Eastman Kodak Company in upstate New York discovered that its film was becoming fogged during shipment. It traced the problem to fallout from US and Russian nuclear tests. At the time, Kodak used material made from corn husks to wrap film shipped in bulk. The corn husks, contaminated by fallout, fogged the film. Kodak threatened to sue over the damaging effects of nuclear fallout on its film. An out-of-court settlement with Kodak provided that the AEC would warn Kodak and other film manufacturers after both US and foreign nuclear tests. After notification of tests, the film companies waited several months before using newly contaminated corn husks for the shipment of film, permitting the radiation levels in the material to drop. The settlement kept the hazards of fallout under wraps, avoiding the public concern and possible jeopardy to the testing program at the continental test site that could have resulted from the publicity surrounding a lawsuit.[12]

On Monday morning, April 27, 1953, the students in Professor Herbert Clark's radiochemistry class at Rensselaer Polytechnic Institute in Troy, New York, were shocked to find that Geiger counters in their lab were registering radiation levels many times above normal. Readings on Geiger counters located closer to outer walls of the lab were the most elevated. Several students took a portable counter outside, finding that wherever they walked, the count rate exceeded normal on the ground, in some places one thousand times the normal readings. Beneath the spouts of rain gutters, the readings were particularly intense.

Dr. Clark hypothesized that fallout from an atmospheric nuclear test at the NPG or PPG or from a Soviet test was the source of the elevated radioactivity. He called John Harley, a friend who worked for the AEC. Clark reported his students' readings to Harley—readings far exceeding the radiation levels he had recorded in the area from earlier nuclear tests. Harley first thought Clark was joking and hung up, then realized there might be something to what the students had found and called him back. Harley made notes on the students' radiation readings, then contacted the director of the AEC's New York laboratory, who was alarmed by the readings and said he would send out some of his people to take measurements and would take any steps necessary to protect public health.

Clark later learned that on April 25, two days before his students measured these high radiation levels, shot Simon had been detonated at the NPG. The resulting mushroom cloud had risen higher than expected, to 44,000 feet above sea level, where winds of 115 miles per hour propelled the fallout to the northeast. The fallout had traveled across the country before being caught in a violent rainstorm over New York state, causing most of the radioactive material to wash out over Washington and Rensselaer Counties, more than 2,500 miles northeast of the test site. Clark's students measured radiation levels throughout the area and found them comparable to levels reported only two hundred to five hundred miles from the test site following atomic tests.[13]

Clark, under contract with the AEC, continued to monitor radiation levels in area reservoirs, while AEC physicists using highly sensitive gamma ray detectors mounted in an airplane conducted surveys of the entire region. The AEC classified the results of these surveys.

Ultimately, the New York State Health Department and the AEC declared the potential health hazards of internal exposure from eating, drinking, or breathing the radioactivity in the area to be negligible. No protective health measures were taken. As testing continued at the NPG during the spring of 1953, rainstorms in the Troy area repeatedly raised radiation levels in the reservoirs to levels equivalent to those measured by Dr. Clark's students following Shot Simon.[14]

Hanford's I-131 Sheep Studies

In early 1950, Hanford workers completed construction on a new Experimental Animal Farm[15] near the F reactor, deep within the interior of the vast Hanford nuclear complex and far from public view. Over the years, up to a thousand animals at a time were housed at the farm and subjected to a range of radiation experiments. Hanford scientists exposed rodents, cats, dogs, pygmy goats, cows, sheep, and other animals to both chronic and acute radiation, often resulting in disability and death for the animals involved.

In the spring of 1950, veterinarian Dr. Leo K. Bustad initiated an

Hanford I-131 sheep studies, Atomic Energy Commission. Photo courtesy of Library of Congress.

experiment referred to as "Problem 690.02"[16] at the new farm to observe health damage to animals from chronic ingestion of I-131. Sheep were chosen for the experiment due to their importance as grazing animals in the Hanford region.[17] Studied over several generations, groups of sheep were fed radioactive pellets containing I-131 at dose levels ranging from .005 to 1,800 microcuries per day (μCi/d).[18] The lowest dose, 0.005 μCi/d, was equivalent to approximately .3 rad/week, the maximum dose recognized as permissible at the time for chronic exposure to whole-body penetrating radiation in humans.[19] A control group received nonradioactive pellets.

The first feeding of I-131 pellets took place in April 1950, when the experimental ewes were well advanced in their pregnancies.[20] Ewes in the highest-dose group, 1,800 μCi/d, became severely impaired after several months, unable to walk, knuckling under, and sometimes remaining prostrate for up to six weeks.[21] After I-131 feeding ended, these ewes died, primarily from pneumonia, or were sacrificed due to ill health before reaching six years of age.[22] Ewes fed 240 µCi/d fared only slightly better. After six to ten months, these ewes became hypothyroid, showing symptoms mimicking those of thyroidectomy. The ewes were lethargic, weak, had trouble walking, and were cold sensitive and very susceptible to respiratory infection.[23] Some of these ewes gave birth prematurely during the second lambing season.[24] Once these ewes became severely hypothyroid, they had little milk for their lambs and showed little interest in them.[25]

In 1951, experiments were begun with yearling ewes fed lower levels of I-131. These ewes experienced lethargy and a milder version of the symptoms seen in the 240 µCi/d and 1,800 µCi/d groups.[26] Ewes fed 135 µCi/d over eight months showed no signs of health damage during the I-131 feeding period. After I-131 feeding was stopped, all but one of the ewes fed 135 µCi/d developed a mild version of the symptoms seen in the ewes fed 240 and 1,800 µCi/d.[27]

During the first lambing season in 1950, Hanford scientists learned that the ewes' milk retained nearly one-third of the daily dose of I-131 ingested;[28] thus, the ewes were passing significant radiation on to their suckling lambs. The effects on first-year lambs born to ewes fed 1,800 and 240 µCi/d were heartbreaking. Lambs born to ewes fed 1,800 µCi/d were weak at birth, unable to stand and nurse, and none survived for more than five days.[29] Lambs born to ewes fed 240 µCi/d had severe thyroid damage, lacked alertness, and weighed less at weaning than lambs in the control group, whose dams had not been fed I-131.[30] These lambs had abnormal development of the skull, with shortened noses and unusually broad heads.[31] An extended lower jaw (prognathia) caused difficulty with chewing following weaning. These lambs failed to gain weight, eventually dying from pneumonia or other secondary infections.[32] Lambs that survived longer than six months had mouth ulcers and gastrointestinal distress and frequently lay down and kicked or walked backward.[33] Lambs born to ewes fed 135 µCi/d were hypothyroid, unable to stand or nurse on their own.[34] They were weak with "stupid and lethargic attitude with drooping ears and depressed head."[35] Even thyroids in lambs of ewes fed 45 µCi/d showed evidence of damage.[36] Lambs from ewes fed 30 and 15 µCi/d showed thyroid damage seven and twenty-two months, respectively, after I-131 feeding was begun.[37]

Second-year lambs of the original ewes fed 1,800 and 240 µCi/d were stillborn or died within four days of birth. Those surviving a few days were very weak and unable to stand to suckle unassisted.[38] Lambs born to ewes in the group fed 240 µCi/d had to receive supplemental feed, as their dams, all severely hypothyroid, had little milk and showed no interest in their offspring.[39]

By 1956, the Hanford sheep studies had revealed that I-131 doses to the thyroid as low as 5 µCi/d for "extended periods" resulted in atrophied, smaller-than-normal thyroids; decreased thyroid activity; and multiple adenomas (benign tumors) on the thyroid.[40] Tumors were not found in the control group fed nonradioactive pellets, indicating that the tumors were likely not part of a natural aging process.[41] Experiments also revealed that thyroid damage was permanent; the thyroid gland did not regenerate or repair itself after I-131 feeding stopped.[42]

The Hanford I-131 sheep data, along with the study findings, were immediately classified by the AEC.[43]

Sheep Deaths in Utah

Operation Upshot Knothole, an eleven-shot series at the NPG detonated March 17 to June 20, 1953, included three particularly "dirty" tests shot from low towers: Nancy on March 24, Simon on April 24, and Harry on May 19. All three deposited especially heavy fallout on southern Utah. Tons of materials from the cabs of the shots and the steel in the towers combined with the relatively low height of the towers (three hundred feet) caused excessive dirt and debris to be sucked up from the test site, contributing to the heavy local fallout. All this material became attached to the radioactive particles and brought the particles down to earth before they could undergo radioactive decay.

The tests released high levels of fallout onto 11,710 sheep grazing in an area from 40 miles north to 160 miles east of the test site.[44] At the time, the sheep were moving from their winter range in an easterly direction toward shearing and lambing grounds near Cedar City, Utah. The AEC did not issue a warning until the fallout cloud had already exposed sheep and sheep ranchers to high levels of radiation. The ranchers saw the nuclear test detonation and watched as the fallout cloud passed over them. Later, according to the ranchers, some people in a jeep came by and warned them to evacuate because "the place was hotter than a $2.00 pistol."[45]

From mid-March through late May 1953,[46] nearly three thousand full-term lambs were born alive and very weak or stillborn, in a stunted condition at approximately half of normal birth weight.[47] Others were born with gross deformities.[48] The lambs that were born alive were so weak that they would lie down and then have difficulty standing. A few tried to nurse but were unable to do so because of weakness or because their dams had little or no milk.[49] Most survived for only five or six days after birth. These symptoms were identical to the health issues seen in the Hanford I-131 sheep experiments in lambs born to ewes fed varying doses of I-131.

Many of the Utah ewes died during lambing or within a few days thereafter.[50] Ewes that did not succumb grew progressively weaker.[51] This outcome was also seen in Hanford experimental ewes fed higher levels of I-131. The Utah ewes had ingested a range of fresh fission products, resulting in beta irradiation of the gastrointestinal tract and, in the process, irradiation of their fetuses. In total, 1,420 lambing ewes (12.1 percent of lambing ewes) and 2,870 new lambs (25.4 percent of new lambs) died during the spring and summer of 1953.[52] Neither the sheep ranchers nor other livestock farmers in the area had ever seen animals affected in this way.[53]

In late May 1953, at the request of a local veterinarian, state and federal veterinarians from the Department of Agriculture visited the ranches. Although they found no evidence of radiation exposure and took no tissue samples, the state veterinarian suggested that fallout exposure might have caused the sheep and lamb deaths.[54] He was concerned that if the sheep deaths were due to radiation, human health might also be at risk from the fallout. He therefore requested aid from the US Public Health Service (USPHS).

The USPHS and the AEC began a joint inquiry in the Cedar City area on June 5, 1953.[55] Ranchers spoke to the investigators about the dead lambs and ewes. The investigators had never seen anything like the deformed lambs and dead ewes that the ranchers reported but did not voice any opinions to the ranchers. Investigators did not examine the sheep and lambs that had exhibited active symptoms and had died, as these carcasses had already been disposed of. The sheep examined were those that had recovered and had symptoms considered mild by the sheep ranchers.[56] Even so, the sheep examined had elevated radiation levels, with radiation found in the spleen, kidneys, ribs, and liver. Beta burns on the backs of many of the sheep and on horses near the NTS were similar to those documented in cattle downwind of the Trinity test.[57]

An AEC investigator's preliminary report identified fallout from atomic tests as a contributing cause in the sheep deaths. The report was confiscated by the investigator's superior, who told him to "rewrite it and eliminate any . . . speculation about radiation damage or effects."[58] The AEC classified the final reports rather than sharing them with Cedar City sheep farmers or local authorities.

Later in the summer of 1953, a second set of AEC investigators blamed malnutrition for the sheep deaths, ignoring the earlier investigators' suggestions that fallout might have been the cause. An AEC official told a state agricultural agent that the AEC didn't want to establish any precedents by paying the sheep owners for their losses.[59] The AEC funded a range study, further promoting the argument that malnutrition due to harsh range conditions was the cause of the deaths.[60]

The AEC's final report found no connection between the exposure of the sheep to radioactive fallout and the deaths of the ewes and lambs. A January 1954 AEC press release referred to studies at Hanford and other AEC sites that had determined that fallout had not contributed to the injuries and deaths of the Utah sheep. Utah health authorities, collaborating with the AEC, issued a final report identifying "unprecedented cold weather, inadequate feeding, unfavorable winter range conditions, and infectious diseases"[61] as the causes of the sheep deaths.

Bulloch I (1956)

The Cedar City ranchers were financially and emotionally devastated by the loss of their herds. Many were forced to take out loans, sell their land, or declare bankruptcy. The deaths of several ranchers were linked to the trauma of losing their flocks.[62]

Unconvinced by the AEC's final report blaming malnutrition and harsh range conditions for the deaths of their herds, Angus Bulloch and other ranchers in Cedar City continued to believe that their sheep had died from exposure to radioactive fallout. In 1954, they filed suit against the United States, referred to as Bulloch I,[63] under the Federal Tort Claims Act (FTCA),[64] asserting $177,000 in damages for the deaths of their sheep and lambs. The ranchers alleged that the government's negligent conduct of atomic testing had resulted in the exposure of their herds to high levels of radioactive fallout, causing death to both sheep and lambs.

Under the FTCA, federal district court judges, sitting without a jury, have exclusive jurisdiction over claims against the United States for injuries caused through the negligence of federal employees acting within the scope of their employment.[65] The sheepherders' case was filed in US district court in Salt Lake City and randomly assigned to Judge A. Sherman Christensen. The government filed a motion to dismiss, arguing that no negligence on the part of government employees had occurred to cause damage to the Utah sheep, and if any damages had resulted from government action, there would be no recovery because atomic testing fell under the discretionary function exception to the FTCA.[66] The court denied the government's motion, contending that the government might still be liable for damages to the sheep ranchers for operational decisions not falling within the discretionary function exception.[67]

At trial, Dr. Bustad, testifying on behalf of the AEC, claimed that the Utah sheep had been exposed to too little radiation for fallout to have caused damage. Bustad had coauthored the report[68] that provided scientific support for the AEC's position that fallout from atomic testing did not cause the Utah sheep deaths. He voiced the opinion that the sheep had died from malnourishment, implying that the sheep ranchers had not taken good care of their animals.[69]

Bustad's report asserted that "radioactive iodine . . . must be administered in quantities in excess of 480 µCi/d to adult sheep in order to produce an impairment of health."[70] Bustad's testimony made no mention of his experimental findings at Hanford, in which the offspring of ewes fed much lower levels of I-131 had suffered thyroid damage and other harm. The report further stated, "The Utah sheep showed no evidence of the radiation damage observed in experimentally treated sheep."[71]

Bustad asserted that the examination of tissues from Cedar City ewes had found no abnormalities except in the liver, indicative of malnutrition or wasting disease.[72] Bustad and his colleagues had not actually examined tissue from dead Cedar City lambs. While he acknowledged this fact, he also reported that no increase in death in neonatal Hanford lambs had been observed in any group where there was an absence of thyroid damage but did not indicate whether the reference to thyroid damage meant the lambs or their ewes.[73] He additionally reported that the large number of deaths seen in the Cedar City sheep had not been seen in any Hanford experimental sheep fed up to 280,000 µCi of I-131 in a single dose, that deaths in the Hanford experimental sheep did not occur until more than five months after I-131 feeding was started, and that those deaths were due to complications of chronic hypothyroidism.[74]

The trial lasted fourteen days. Experts failed to link the sheep deaths to fallout. As a consequence, in October 1956, Judge Christensen ruled against David C. Bulloch, the bellwether plaintiff representing the group of sheep ranchers. The court, while recognizing that the AEC had been negligent in not warning the sheepherders about fallout that would be deposited in the area in advance of the atomic test, ruled that because the deaths of the sheep were not caused by radiation released from atomic testing, negligence by the government had not proximately caused the sheep deaths, and the plaintiffs were not entitled to recovery.[75] As the result of this ruling, the government did not have to hide behind the discretionary function exception to avoid liability. Bulloch and the other sheep ranchers did not appeal the decision.

The "Beautiful Bomb"

The uninterrupted atmospheric testing of nuclear weapons within the continental United States required the unwavering support of the public. The public had grown increasingly concerned over the safety of the tests, with mushroom clouds on full Technicolor display just over the horizon and fallout clouds blanketing communities downwind. The AEC embarked on an aggressive campaign to reassure the public that the tests posed no risk, encouraging people to bring their families to witness the tests.

In January 1955, the AEC distributed a booklet entitled *Atomic Test Effects in the Nevada Test Site Region* to communities downwind of the test site.[76] The booklet, written in grade-school-level text, featured a cartoon of a group of smiling, seemingly carefree men, women, and children framed by huge blast clouds and a drawing of a man and woman wearing sunglasses, dressed in Western garb, smiling and waving as a massive cloud expanded behind them. The booklet claimed, "No one outside the test site in the nearby region of potential exposure has been hurt," and fallout "does not constitute a serious hazard to any living thing outside the test site."[77] The AEC pacified downwind communities with publications and images depicting atomic tests as harmless. The mushroom cloud was not to be feared; its "mass of smoke rising thousands of meters in the air, swirling in strange, lovely colors like the rainbow"[78] before expanding into a mushroom cap in the stratosphere was merely a form of spectacular family entertainment. As a nuclear historian at the Hiroshima Peace Institute observed, "The bomb presented to these specific communities was a beautiful bomb, a Disney bomb."[79]

One indication of the effectiveness of the AEC's public relations campaign was the exuberant response of the Chamber of Commerce and a number of casinos in Las Vegas, close enough to the test site that the mushroom clouds were clearly visible. The city capitalized on the entertainment value of the mushroom clouds: calendars printed by the Chamber of Commerce advertised detonation times and the best locations for observation of the tests. Casinos offered special atomic cocktails and "dawn atomic parties" with dancing and drinking on casino rooftops until an atomic flash filled the predawn horizon to the northwest. Women wearing mushroom clouds sewn over the front of their swimsuits competed for the title of "Miss Atomic Bomb" at the Sands.[80] One casino magnate reported, "The best thing to happen to Vegas was the Atomic Bomb."[81]

In spite of the AEC's repeated assurances that the atomic tests posed no danger, members of downwind communities began to carry Geiger counters to measure radiation levels following the detonations.[82] Local news reported stories of Geiger "counters going 'off the scale' after test shots." The AEC told the public that off-the-scale Geiger counter readings following atomic tests were nothing to worry about.[83]

A Geiger counter was brought along in the summer of 1954 for the filming of the big-budget Hollywood film *The Conqueror* in Snow Canyon State Park, eleven miles northwest of St. George, Utah. The film starred John Wayne as the twelfth-century Mongolian warlord Genghis Khan and Susan Hayward as a beautiful captive princess. Director Dick Powell had chosen the area for its resemblance to the steppes of central Asia. St. George, approximately 213 miles due east of the test site, had been plastered with particularly high levels of fallout from a number of the tests at the NPG, now renamed the Nevada Test Site. During the filming, St. George found itself overrun with actors, producers, technicians, and stuntmen. Producer Howard Hughes, eccentric millionaire and head of RKO Pictures, poured vast amounts of money into the project.

Huge fans powered by generators were set up in the desert to create Hollywood-style sandstorms. Sword-wielding actors fell off horses, swallowing mouthfuls of radioactive dirt in the process. Extras, engaged in hand-to-hand combat, rolled through the red dust. Everyone ate fallout-contaminated food. Some of the actors and crew were aware of the radiation danger; a Geiger counter set up near the film set crackled so loudly that Wayne thought it might be broken. But AEC officials had reassured everyone that the area was completely safe, so the insistent warnings of the Geiger counter were dismissed. When the majority of filming in Utah was complete, sixty tons of radioactive dirt were shipped to Hollywood to re-create the scenery for final filming on sound stages.

The film was a gigantic flop, but much worse was to come. Many members of the cast and crew may have paid the ultimate personal price for participating in the project, filmed in an intensely radioactive environment that bestowed significant exposure on everyone involved. In the years following the filming of The Conqueror, of 220 cast and crew members, 91 developed cancer, and many died from the disease, including stars John Wayne (stomach cancer) and Susan Hayward (brain cancer), as well as Agnes Moorehead (uterine cancer) and director Dick Powell (lymphatic cancer). Another actor, Pedro Armendáriz, survived kidney cancer but then committed suicide when he learned that he had terminal cancer of the lymphatic system. Approximately three hundred Shivwit Paiute Indians, playing Mongol villagers, sat around in the dust between takes. It is not known whether any of the Shivwit Indians went on to develop cancer.[84]

Testing Continues in the Face of Public Protest

Bulloch I had brought the potential health effects of fallout exposure on animals and, possibly, on humans, into the public eye. Growing numbers of antinuclear protests took place at the gates of the NTS and around the world. On April 23, 1957, Dr. Albert Schweitzer, the French German physician, humanitarian, writer, and philosopher, publicly called for an end to atmospheric nuclear tests.[85] His "Declaration of Conscience," broadcast worldwide, emphasized the dangers of fallout from nuclear testing. Schweitzer's declaration helped catalyze the growing public concern about the nuclear arms race.

By 1957, atmospheric testing at the continental test site had entered its sixth year. For the first time, film badges measuring external radiation exposure were provided by the AEC to a few civilians in geographically widespread locations downwind. The AEC continued to closely monitor its test site personnel through the use of individual film badges, radiation exposure diaries, pocket dosimeters, and other instruments. The AEC also regularly recorded on-the-job radiation exposure of employees in its national labs. National laboratory employees handled quantities of radioactive substances representing only a tiny percentage of the levels of radiation contained in the fallout released downwind from atomic tests.

In spite of mounting public protest, atmospheric testing at the NTS continued unabated. Operation Plumbbob, May 28 through October 22, 1957, was the sixth test series at the NTS. Plumbbob, a series of twenty-nine shots, included several that were part of Desert Rock exercises to assess troop readiness on the nuclear battlefield. Desert Rock shot Smoky was a forty-four-kiloton shot fired from a seven-hundred-foot tower on August 31. Troop exercises near ground zero immediately after detonation exposed over three thousand servicemen to high levels of radiation.[86] Mortality and the incidence of leukemia in these exposed atomic veterans would not be studied until 1979.[87]

On August 6, 1957, a group of Quakers, Mennonites, and other pacifists traveled to the test site to observe the twelfth anniversary of the 1945 atomic bombing of Hiroshima and to protest the continuation of atomic testing. The protesters camped close to the boundary of the NTS and held a vigil during Plumbbob shot Stokes. Those crossing onto the test site were arrested.

Also in the summer of 1957, Congress held the first full congressional hearings on the health effects of fallout from atomic tests.[88] Sponsored by the Joint Committee on Atomic Energy, the hearings allowed the AEC to tell its version of the facts about the health hazards of the combined global fallout from British, American, and Soviet atomic and thermonuclear tests.[89] The committee convened hearings several times over the next seven years until atomic tests moved underground and out of the public eye in 1963 with the signing of the Limited Test Ban Treaty.[90] Once tests went underground, concerns over atomic testing vanished from the public conscience in spite of the fact that a number of underground tests, including the Baneberry test on December 18, 1970, vented significant amounts of radioactive fallout downwind.[91]

5

I-131, a radioactive isotope of iodine, was one of the fission products released in fallout from atomic testing and from the chemical separation of plutonium at Hanford. When the I-131 was deposited on pasture grass, it was consumed by grazing cows and goats. It then passed into the animals' milk and was ingested by humans, either as fresh milk or after pasteurization and processing. This I-131 pathway from fallout to pasture to milk to humans is referred to as the "milk-iodine pathway."

I-131 has a half-life of eight days, referring to the amount of time it takes for its radioactivity to fall to one-half of its original value. If milk is consumed fresh, rather than after processing in a dairy, the I-131 levels in the milk are significantly higher because in fresh milk, I-131 does not have time to undergo radioactive decay before it is consumed. The human thyroid absorbs I-131 from the milk, just as it would normally absorb nonradioactive iodine, which it requires in order to produce thyroid hormone. Children are particularly vulnerable to the effects of I-131 both because children's thyroids are still developing and because children tend to consume large quantities of milk and other dairy products.

On October 10, 1957, at the Windscale plutonium production facility in northwest Britain along the Irish Sea, a reactor accident resulted in a fire that released radiation off-site for several days. About twenty thousand curies of airborne I-131 escaped from Windscale. Local milk was sampled and found to be contaminated with I-131. Milk cows downwind of Windscale were quarantined, and thousands of gallons of contaminated milk were confiscated by health authorities and then dumped into the Irish Sea.[1]

Because British health officials knew to test the local milk supply for I-131 at the time of the Windscale accident in 1957, it is clear that they were aware of the milk-iodine pathway. Scientists were in fact aware of the uptake of iodine into milk as early as 1919.[2] The earliest studies were done with the natural form of iodine, as the radioactive isotopes of iodine, including I-131, had not yet been discovered. An article published in 1930 reported that feeding iodine compounds to cows increased the iodine content of their milk.[3]

By the early 1940s, the scientific community was aware that radiation could cause cancer. By 1941, I-131 was known to concentrate in the thyroid, and by 1942, scientists knew that hypothyroidism could follow radiation treatment of the thyroid.[4] Manhattan Project scientists reported in March 1943 that releases of I-131 could cause damage by concentrating the radioiodine in the thyroid.[5] By November 1943, I-131 in the thyroid was known to cause thyroid destruction.[6] Prior to Hanford's start-up and long before atmospheric atomic tests began in 1951 at the continental test site, the scientific community understood that I-131 could be passed through radiation-contaminated milk to humans and that that I-131 could, in turn, cause damage or destruction of the thyroid.

Hanford's off-site release of I-131 and other radioactive isotopes began with the start-up of the facility in late 1944. Dr. Joseph Hamilton at the Lawrence Berkeley National Laboratory had conducted early work for the Manhattan Project with inert (nonradioactive) iodine and its radioactive isotope, I-131. Hamilton had suggested that the preventive feeding of inert iodine would block the harmful effects of I-131 on the thyroid.[7] Manhattan Project scientists also advised that the administration of potassium iodide pills could protect the thyroid from I-131.[8] In 1945, a secret Hanford memo referenced the addition of iodine to the drinking water supply in Rochester, New York, to prevent goiter (a thyroid disease) and advocated that the Hanford-area public be advised to use iodized salt or that only iodized salt be made available for sale.[9] Hanford officials also proposed that a dose of iodine be given to Hanford chemical separations workers during their monthly medical examinations.[10] However, this preventive measure would have alerted workers that they were being exposed to hazardous levels of radioactive gases. "This plan was never completed. . . . The security aspect of iodine feeding to groups not acquainted with the problem was considerable. Questions relative to the need . . . are difficult to answer without provoking some undue alarm."[11]

An article in the *Richland Villager* in 1954 promoted iodized table salt for "good nutrition." Residents of Richland were told that the very visible testing of local milk taking place in town was for the purpose of measuring bacterial count and butterfat content, to ensure that the milk came from tuberculin-tested cows, and to confirm that proper pasteurization had been carried out.[12]

In 1951, the sheep experiments conducted at the Hanford experimental farm revealed that a large percentage of the I-131 fed to the sheep ended up in the sheep's milk. Hanford experiments on cows found a similar transfer of I-131 to cows' milk, leading to the conclusion "that I-131 would be found in such dairy products as skim milk, cottage cheese, and whey."[13] In 1955, Hanford's director of radiological science confirmed, without informing the public, that scientists had known since the first year of Hanford operations that permissible levels of I-131 in the Hanford region were too high and that "drinking of milk from cows on contaminated pastures," and not inhalation, was the problem.[14] Hanford scientists tried to control the risks of I-131 by establishing a maximum safety level of one rad per day of I-131 exposure for members of downwind communities.[15] People living downwind, however, were subjected to continuous exposure, resulting in elevated cumulative radiation exposures even when the exposures were limited to the maximum safety level of one rad per day of I-131. Releases, however, sometimes exceeded the safety levels by two to two and a half times.[16]

In spite of the long-recognized milk-iodine pathway, the AEC failed to notify populations downwind of Hanford and the test site of this hazard, did nothing to remediate the risk, and did not confiscate contaminated milk supplies to lessen the danger to public health. Following NTS shot Harry on May 19, 1953, "radiation monitors decided against taking milk samples in order to avoid arousing public concern."[17] A USPHS official in St. George, Utah, "considered collecting milk samples from local dairies to check for radioactivity, but because of the uneasiness in the community concluded that such a survey might create alarm."[18]

Neither the AEC nor the USPHS measured internal exposure to

radioactive iodine downwind of the NTS through thyroid or milk measurements until 1957. There was no systematic collection or radiochemical analysis of milk and locally grown foods during years of atomic testing.[19] Even when indirect estimates of internal exposures were made through monitoring milk or food, these efforts were haphazard at best.[20]

In the early 1960s, a scientist with the AEC wrote a report criticizing the AEC's apparent ignorance of the introduction of I-131 from fallout into the milk supply in downwind communities.[21] The AEC initially refused to publish the report, dismissing the work as inadequate[22] and eventually classifying the document as a confidential report. It was not released until 1978, in response to Freedom of Information Act requests.

It is difficult to know how many children downwind of Hanford and the NTS went on to develop thyroid cancer following exposure to airborne I-131. Thyroid cancer is sometimes referred to as the "good cancer"[23] because of high survival rates in people whose cancer is caught early. Unfortunately, even if people survive thyroid cancer, they often find themselves burdened with significant lifelong disability resulting from thyroidectomy and the combined impact of chemotherapy and radiation treatments.

Michael Helland and Dan S. grew up downwind of Hanford. Both survived thyroid cancer that they believe to be the result of childhood exposure to Hanford's I-131. Both now cope with disabling and potentially life-threatening health issues as the direct result of aggressive cancer treatments.

Here are their stories.

Plaintiff 11: Michael Helland

I was born in 1946 in the Spokane Valley, east of Spokane. Any radioactive fallout that came across from Hanford was deposited in the Spokane Valley. That fallout pattern was easily visible when ash clouds traveled from the Mount St. Helens volcanic explosion in western Washington in 1980, and large amounts of volcanic ash concentrated in the Spokane Valley.

I was only twenty years old when they discovered papillary thyroid cancer. I was home following my sophomore year at Whitworth College, about to work for the Forest Service for the summer. I complained to my mother about a sore throat and fatigue, and she noticed some swelling in my neck. She thought it looked like a goiter, a swelling of the thyroid. She

Michael Helland and his mother. Photo courtesy of Helland family.

was familiar with goiters because she was from the Midwest, where goiters were common, particularly in my great-grandmother's generation, before they figured out people needed iodine. Now there is iodine in the salt so people don't get goiters.

We went to the doctor, and he told us that this kind of thing really wasn't good in someone my age. He sent me to Spokane's only endocrinologist, who told us when a person is eighteen to twenty years old, metabolism changes. If there is toxic exposure earlier in life, thyroid cancer often shows up at about this time. He didn't know then about the radiation releases from Hanford, since that information wasn't made public until 1986, and this was 1966. But I will remember a comment he made for the rest of my life. He said they were seeing a lot of thyroid cancer in the young people in the area. That is what was seen after the Chernobyl accident too.

The medical profession had just started using I-131 scans back then, and the procedure was ordered for me. When they did the I-131 scan, I could hear the Geiger counter go off when they passed it over my thyroid. It went off every time the radiation scanner passed over me. I'm feeling like I probably got a pretty big dose of I-131 based on that scan. The scan revealed that all twenty-five to thirty lymph glands in my neck had been invaded by malignant thyroid tissue. This was truly frightening, particularly to someone as young as I was at the time.

They tried to kill my thyroid cancer three ways—through radical neck surgery to remove the cancerous nodules and thyroid, then through massive cobalt irradiation of my neck, and finally with I-131 ablation treatment.

I had the radical neck surgery shortly after the I-131 scan revealed that thyroid cancer had metastasized throughout all the lymph nodes in my neck. My thyroid and all the lymph nodes in my neck were surgically removed. The surgeon basically opened me up like a sardine can, peeled back my skin, and took everything out. The surgery destroyed my parathyroid glands, which regulate blood calcium levels, so now I have hypoparathyroidism. I didn't know it at the time, but hypoparathyroidism would be a very challenging health issue for me for the rest of my life.

The parathyroid glands, tiny glands adjacent to the thyroid, control blood calcium levels. After the surgery, because my parathyroid glands were destroyed, my calcium levels would plunge dangerously, causing involuntary muscle spasms called tetany. In the hospital following surgery, I woke up from a nap and found my arms spasming really hard, so I called the nurse on the call button. No one responded. I couldn't yell because my vocal cords were messed up from the surgery, so I started whistling loudly. A nurse tracked down where the whistling was coming from. When she saw me spasming, she ran to get help. They had to do an immediate intravenous calcium infusion to get me stabilized.

Shortly after the surgery, I started choking, and they nearly had to perform an emergency tracheotomy, which would have required cutting a hole in the front of my neck to put a tube into my trachea so I could breathe. Fortunately, I somehow started to breathe again on my own, avoiding the tracheotomy, but it was close. I continue to have a lot of problems with swallowing and occasional choking due to scar tissue impinging on my esophagus as the result of the radical neck surgery.

The surgery caused nerve deterioration where the nerve ends were cut in my shoulders. I could no longer play golf because I couldn't tell I had hit the golf ball. That was a big loss to me, as I really enjoyed the game.

Due to my hypoparathyroidism, I wasn't able to maintain high enough blood calcium levels for the first ten years after surgery until dihydrotachysterol (DHT), a vitamin D derivative, was developed. The strength of the DHT, however, was unreliable, so it was very hard to regulate my vitamin D levels. Before DHT, regulating vitamin D levels was even more difficult. I accidentally overdosed on DHT in 1988, resulting in kidney failure from extremely high calcium levels. I suffered permanent damage to my kidneys.

After the surgery, I ended up with a very disfiguring fifteen-inch burgundy-colored scar from ear to ear with keloids [scar tissue] on both sides. The scar was so noticeable for the first year or so that I always felt like people were looking at me, trying to figure out what was wrong with me. I was still really young, and this was very hard on me.

After the radical neck surgery, I had to go through cobalt irradiation. I lay down under a radioactive emitter, and they put lead-lined shields around the areas they didn't want irradiated. They repeated the irradiation for several weeks. It caused hair loss and blanching of my skin. They told me to keep the irradiated areas out of direct sunlight. They wouldn't subject a person to that much high-level radiation today.

The third thing they did, which is what they do now, was to give me a massive dose of I-131 as ablation therapy to burn up any remaining thyroid tissue. During ablation treatment, I had to stay in a lead-lined hospital room, and I couldn't come within ten feet of anyone. In those days, the ability to minimize the dose and to regulate this kind of treatment was primitive. It was a Catholic hospital, and the only one who would come into the room was an elderly nun—none of the younger people would come in. I remember someone told me, "You shouldn't worry; you aren't going to have kids anyway!" How did they know that?

My parents came to visit me in the hospital after the I-131 treatment, and they had to stand fifteen feet from the bed to talk to me. I had to flush the toilet twice to flush down any radioactivity. It was scary and kind of eerie, to say the least.

After ten weeks of feeling truly lousy because I wasn't allowed to take thyroid medication, a scan said I was finally cancer-free.

I started having symptoms of ankylosing spondylitis within a few months of going through the radiation treatment. Ankylosing spondylitis causes vertebrae in the spine to fuse. I also suffered from recurring sciatica, involving nerve pain in the back of the upper legs, every three to four weeks, traveling down one or the other leg, causing extreme pain and drastic limping. Thank God it never hit both sides at once. One doctor later told me that the radiation treatment most likely accelerated the condition and possibly worsened the symptoms because it's an autoimmune disease.

When the ankylosing spondylitis symptoms started to appear, I didn't even want to find out what this new problem was. I had enough issues with the thyroid cancer and the effects of treatment. I felt, one more mark on my record and no one would ever hire me! The ankylosing spondylitis wasn't diagnosed until I was thirty years old. I just hurt, but I didn't know what it was. Changes in barometric pressure made a big difference in how I felt. Every time a low front came through, it would get worse, then flare for several days.

I have to take synthetic thyroid for the rest of my life, and it has always been a battle to keep both the thyroid hormone and calcium blood levels regulated correctly.

My personality totally changed as the result of everything I had gone through. I became much less outgoing. I now needed to be on a path with a defined goal. I didn't have those young-person dreams of time off, taking trips. I matured far too quickly.

I am convinced my thyroid cancer was caused by my childhood exposure to airborne radiation from Hanford. It is known that Hanford's fallout reached Spokane, sometimes forming a hot spot. My grandmother was big on natural foods, so we got fresh unpasteurized milk. We also had fresh vegetables grown in our garden that would have been contaminated as well.

If you live in Spokane, you know that in the summertime, there are at least one or two dust storms blowing in from the direction of Hanford. Sometimes, back when I was growing up, huge dust storms would race across the Columbia Plateau. You would see the whole sky turn dark orange, and you would know one was coming. The vegetation is much thicker nowadays because of irrigation and the many wineries, but in those days, that area was barren desert. And, if you read about Hanford back then, you'll see mention of "termination winds," major windstorms that kicked up the desert sands and made Hanford workers want to leave the area. I read about one windstorm that caused three thousand people at Hanford to leave in one day! They used to say, "On a clear day, you could see across the street." I think the huge dust and windstorms were common back then.

My younger sister's health may also have been impacted by exposure to Hanford radiation. My sister had fibromyalgia, which was not a diagnosed condition back then. She was always in pain. One time, I took her to the doctor's office, and the doctor showed me her leg, salmon pink with silver-dollar white spots on it. He said that these were the signs of people who are in extreme pain. I knew for certain then that her pain was real and not something just in her head.

I moved to Seattle to attend the University of Washington after my thyroid cancer treatments. My scar brought a lot of unwanted attention. I was very self-conscious about it, as the scar was long and bright burgundy. I stayed about eight years in Seattle, moved to Boston for a couple of years to go to school, then to Chicago and Houston, working in oil exploration.

I returned to Spokane in 1993, having made the decision to retire due to the health problems and the constant pain. It really wears you down. When I returned to Spokane, I learned about the radiation releases from Hanford and began to make the connection to my childhood exposures.

I continue to have health issues caused by thyroid cancer and its treatment. These problems are lifelong and are worsening as I get older. I have constant pain in my neck and upper back from two completely degenerated neck discs in the area that received cobalt irradiation. The lack of lymph glands in my neck adversely affects my recovery from viruses and sinus infections. Ankylosing spondylitis is an inflammatory disease that can cause vertebrae in the spine to fuse together. Eventually, spurs on the vertebrae grow together, and the spine becomes rigid as if it had been surgically bonded. It starts in the lower back, then it moves up the spine over time. It's like a severe arthritis. The progression of vertebral fusing has somewhat helped the pain in my lower back, as the spine is now rigid in that area, but the pain up higher is intense.

Ankylosing spondylitis calcifies the spine, causing severe stiffness. It's really bad in the morning, and it takes a while to limber up. People don't understand why I'm a night owl. It's because it takes me all day to start to feel better. It's lucky I do the exercises I was given by a physical therapist early on, as they have helped maintain my range of motion over the years. I know, though, that I am becoming more limited as I age. Where once, lying on my stomach, I could lift my head, neck, and shoulders almost eighteen inches off the floor, I am now down to only one inch. Had I not done those exercises religiously, I am convinced I would now be bent over forward from spinal deformity.

I started on a medication for the ankylosing spondylitis when I was thirty years old, and it began to eat my stomach lining. Other meds also had bad side effects. Now I just take massive doses of aspirin. I also have a hot tub at home that helps.

I'm very concerned that my neck might break if I ever have to be intubated for a medical procedure. My neck can't be bent back due to my condition. It's a life-threatening issue.

A malignant melanoma was found on my neck in the area that was irradiated for several weeks during cobalt radiation therapy. The melanoma was removed in 2000. I just had another suspicious lesion biopsied, and I am waiting for the pathology results. It's a constant concern.

In 2004, my TSH numbers were starting to be too high, which means I was hypothyroid. It was also the first indication that the thyroid cancer was back. Blood tests and an I-131 scan confirmed recurrence of thyroid cancer with several nodules in both lungs, the largest of which were pea-sized. I was once again treated with massive doses of I-131 radiation. I was hospitalized and isolated until the radiation dropped to safe levels. I-131 scans after both treatments showed several small areas of thyroid cancer still remaining in my lungs.

I have reduced lung capacity due to cancer and radiation scarring, enough so I am out of breath pretty quickly. I used to sing in a choir, but I can't keep up anymore, as I end up gasping for breath.

There are now biologics that might help with the ankylosing spondylitis. But if you have thyroid cancer, you can't take biologics, as they suppress the immune system, and it is the immune system that keeps the thyroid cancer from growing.

In 2007, a large protuberance showed up at the juncture of my collarbone and sternum, causing concern that my thyroid cancer may have spread to the bone. After two MRIs, a bone scan, and an X-ray costing $2,400 after insurance, it was diagnosed as a bony effusion related to arthritis.

In 2010, a thyroglobulin test indicated the continued presence of active thyroid cancer. My doctor is concerned that the cancer will spread to my brain.

I have struggled for almost fifty years to maintain proper thyroid replacement hormone and blood calcium levels. In 2012, insufficient thyroid replacement medication caused me to be lethargic and gain weight for almost nine months until my thyroid hormone levels were again regulated correctly. I feel like I slept away nine months of my life before my doctor figured out what was wrong. Then, I had below-normal blood calcium in 2013 due to insufficient calcium and vitamin D levels. I am always having to argue with my insurance provider over coverage for vitamin D derivatives that I require, costing $150 per month—their argument is that "vitamins are not covered."

I have to maintain high-normal levels of thyroid replacement hormone to reduce the chance of further thyroid cancer growth. It's like a feedback loop. The high levels of thyroid hormone cause me to be irritable and heat intolerant. I need expensive Thyrogen and thyroglobulin tests regularly to monitor thyroid cancer tissue growth. I can only hope that, with thyroid hormone levels kept high, my TSH levels are kept correspondingly low, preventing the cancerous thyroid masses in my lungs from growing.

I am tired a lot of the time. It depends on whether I have too much thyroid medication or too little.

My doctors are just watching the remaining thyroid cancer in my lungs. There are sophisticated tests now, called TSH-stimulated (Thyrogen) thyroglobulin tests, that can measure metabolic waste produced by thyroid cancer cells. You have to go off your medication for a week, which is really uncomfortable, during these tests. It makes you depressed and rundown.

These health issues, which I attribute to childhood exposure to Hanford airborne radiation, have caused major interruption and detriment in my life from the time I was young.

I always think, "What next?" People who haven't gone through this don't get it. These health issues played a role in the end of my first marriage. When my former wife wanted to have fun or go out after work, I never had the energy to do anything.

Michael Helland. Photo by author.

I feel dysfunctional because I have to worry about all these problems all the time, but if I don't pay attention to them, they get out of whack. I don't try to do anything rigorous late in the day. People probably get the wrong impression when I get up and leave gatherings after an hour or two because I feel uncomfortable. Only one or two people in my church know I have ankylosing spondylitis. I just don't want that to be the focus.

Otherwise, a person more or less becomes their disease. But it's getting so severe that I will probably have to tell people what's going on.

The main thing I think people should know about Hanford is that the government strongly opposed our personal injury actions, stringing out the Hanford downwinder litigation for years when they knew that we had valid claims. I don't think we thought it would ever be resolved. Tom Foulds, my attorney, is a hero. He was out so much legal expense over the years without compensation, yet he kept urging us to hang in there.

The defense had no incentive to settle—defense attorneys had endless dollars to spend under indemnification agreements. They spent ten times as much fighting our claims as they spent settling with injured downwinders. They were afraid that otherwise, legal precedent would be set.

Plaintiff 12: Dan S.

I was born in May 1950. My family lived in a small town with no hospital, so my mother gave birth to me at Deaconess Hospital in Spokane, about twenty-five miles away.

We lived in Lincoln County, approximately 146 miles northeast of Hanford, until I was three years old. We then moved to Wenatchee, about 106 miles northwest of Hanford, until my family relocated when I was four years old to Douglas County, around 140 miles to the north of Hanford. I remained in Douglas County through graduation from high school in 1968. Following graduation, I went to Washington State University in Pullman for four years, with summers in either Whitman or Douglas County. I left eastern Washington in 1972, returning two years later to the town of Grand Coulee, about 140 miles north-northeast of Hanford, where I lived until June 1976.

Altogether, I spent nearly twenty-three years in towns downwind of Hanford; first in utero, then as an infant, child, and young adult. I was in the womb in Lincoln County during the Hanford Green Run experiment in December 1949.

Some of the time we got our milk straight from the cow. I know this because I have an old letter from my dad where he described milking the backyard cow during my childhood. Fresh milk can contain significantly higher levels of radioactive iodine than milk first processed in a dairy.

My dad always kept a vegetable garden, and we got all our fruit locally.

Dan S. as a child. Photo courtesy of Dan S.

Both vegetables and fruit could have been contaminated by airborne radiation that was deposited on everything.

As I got older, I started to have health issues. By the time I was thirty-eight, I knew something was definitely wrong with me. I was very tired all the time. I kept gaining weight even though I changed how I was eating. I lacked energy. I had dry skin, hair loss on my legs and scalp, and areas of my skin with loss of pigmentation, which I now know are caused by hypothyroidism. My mind wasn't 100 percent either.

I moved to the Tri-Cities in 1990 and saw the first articles on Hanford and the downwinders. I had never heard of "downwinders" before. I read the stuff with fascination.

In 2000, my internist found nodules on my thyroid. A fine-needle biopsy revealed the nodules were suspicious for cancer. My physician ordered a dye test, and the trachea was highlighted, meaning that multiple nodules had grown in my thyroid. That explained why I was having a lot of trouble swallowing and sometimes choked on my food. The nodules were pushing in on my trachea.

The dye test was followed up with a whole-body scan to see whether the cancer had metastasized elsewhere in my body. Not much showed up anywhere else, thank goodness. I was really worried that the cancer would completely destroy my trachea. I thought at the worst, I could at least breathe through a tracheostomy hole cut into my neck if that became necessary. I wanted to survive.

I had a subtotal thyroidectomy in July 2000, sparing my parathyroid glands. The surgeon who took my thyroid out told me it looked "clean," that the cancer most likely hadn't metastasized outside of the containment capsule, so I thought that was the end of that.

I then went through radioiodine ablation therapy followed by 5,400 rads of nuclear-accelerated X-ray radiation at the Tri-Cities Cancer Center. My radiation oncologist asked me whether I had ever lived in a "downwinder" county. I learned that my form of thyroid cancer could possibly be linked to I-131, one of the radioactive substances released from Hanford.

I began to read about the counties in Washington State impacted by Hanford fallout and realized that during the period I was in the womb and for the first twenty-two years of my life, I was exposed. My exposure as a fetus was probably the most significant, as my thyroid was just forming.

My mother likely drank a lot of I-131-contaminated milk during her pregnancy, and that radioiodine passed across the placental barrier to me.

Following the thyroidectomy and until I finished a second radioiodine ablation treatment session, there were four months when I was without any thyroid or thyroid hormone in my body. I had no energy and no energy reserve. Simple tasks took forever. Just trying to mow the lawn, which would normally take forty-five minutes, would take three times that long, with rest breaks. I also had pressure in the head, with a "grainy" feeling.

Because I had stage 2 papillary thyroid carcinoma and was age fifty, the protocol was to do six weeks, five days per week, of nuclear accelerated X-ray radiation. All of my major medical problems come from that extensive radiation treatment.

I wore a face shield during the radiation sessions, so I didn't lose my teeth, but I am no longer able to grow a beard in the area on my neck that wasn't covered by the shield. Now, when I move my head or jaw, there is a popping, clicking noise due to hardened, nonflexible tissue caused by the repeated irradiation. I have also had chronic dry mouth ever since the treatment. Dry mouth has negatively affected my teeth. My digestion is also problematic now. I'm not sure why I have digestion issues, as the radiation therapy wasn't aimed at that area of my body.

The radiation therapy burned my skin. I have to watch for melanomas developing within the irradiated areas. They just removed one precancerous lesion, and I'm waiting for the pathology report. I'm concerned about another suspicious lesion as well.

Chemotherapy has been proven to predispose one to adult-onset diabetes, and radiation is now presumed to cause that predisposition as well. I was diagnosed with adult-onset diabetes when my A1C blood level reached 6.9. I am trying to control the diabetes with diet. None of my siblings have diabetes, so I'm not sure why it hit me. I'm thinking it was the radiation and ablation therapies that caused it.

As a result of the radiation treatments, I now have dysphagia, which involves a lot of difficulty swallowing. In order to take a vitamin pill, for instance, I have to cut it in many small pieces. Everything I eat has to be followed by or simultaneously swallowed with liquid. I also have odynophagia and globus, which is a sensation of a lump in my throat. If I have hot food or liquid or have to vomit, my throat really hurts afterward.

The radiation scarred my lungs, and the radiation oncologist would occasionally pan further down and hit the top of the pancreas in case there were any cancer metastases there. I don't know if any damage was done to the pancreas.

I also have "feline esophagus" as the result of the radiation therapy, as my esophagus is now similar in structure to that of a cat. For cats, food and fur can go up or down, either direction. Cats have at least seven concentric rings in the esophagus. I went to a gastroenterologist to get him to open up my esophagus, as it is so tight and because swallowing is a huge problem, but he was worried that any such procedure would damage and tear tissue, and so he said he would only do it as a last resort.

When they irradiated me, it damaged the sphincter at the top of the esophagus, so I now have GERD. The damaged sphincter allows backwash of stomach acids. GERD produces epigastric and chest pain, heartburn, nausea, vomiting, and regurgitation.

I'm in trouble if I'm eating and I don't have water handy. This is a particular problem in a restaurant if the server doesn't constantly refill my water glass. I can swallow soft liquid stuff, but any more solid food is a problem.

As a result of the radiation treatment, my throat is a mass of fibrosed tissue. If I get food stuck, it covers the epiglottis and I can't get any air into my lungs. I can't call for help. I quickly turn blue. I'm apt to die while waiting for someone to perform a Heimlich maneuver.

I have severe vocal cord damage from the radiation treatment. The doctor recently did an endoscopy to look at my vocal cords. The vocal cords should be thick and flesh colored, but mine are bright red and thin, with spider veins. My voice changed from the radiation, to the point where my little sister didn't recognize me over the phone. I had to convince her that it was really me on the phone by sharing anecdotes from childhood that only she and I would know. My voice fades now after just a few hours of speaking. I am hoarse a lot. I can no longer sing, and I cannot raise my voice to get someone's attention.

Sometime after my cancer diagnosis and treatment, I saw an ad in the local paper from a law firm, alerting people with thyroid disease or cancer who had lived in certain counties that they may have been exposed to airborne radiation from Hanford. I gave them a call and was added as a plaintiff to the Hanford downwinder litigation.

When I went in for my five-year checkup after thyroid cancer treatment, they did a radioactive dye test and, unfortunately, found something in my hip. My doctor told me to go to the hospital right away to get an MRI and X-rays. I had twenty-one X-rays! I was worried that the thyroid cancer had now metastasized to my hip. I started wondering how much longer I would be alive.

A team of physicians looked at the MRI and X-rays and decided it was an enchondroma. I didn't know what that was. Turns out an enchondroma is a noncancerous bone tumor that begins in cartilage. It had apparently been there since I was young. It was just big enough to show up. They had thought it was cancer, so they had X-rayed me twenty-one times. Think of all the excess radiation I got needlessly! I already had a lifetime of radiation from the thyroid cancer treatment, and now this.

I am very tired all the time now. I take levothyroxine synthetic thyroid hormone daily. My doctor keeps cutting back my dose. I need more energy, but he thinks my heart rate is already too high, and I was too jittery at higher thyroid hormone doses. If I don't take a nap regularly, I can't handle it. I also feel cold all the time, which is a symptom of hypothyroidism.

My immune system is weak as well. I get sick faster than other people around me.

I have disseminated idiopathic skeletal hypertrophy, which is hypercalcification. It may be due to hyperparathyroidism, with the parathyroids putting out too much parathyroid hormone, resulting in too much calcium in my blood. It's possible my parathyroids were damaged when I was exposed as a child, but I will never know for sure.

Recently, adding to my ever-longer list of health problems, I was diagnosed with prostate cancer.

My mother and her mother also had some health issues that may have been related to Hanford's radiation. My mother survived a melanoma on her leg in 1948 in Chelan County, and her mother, who lived in Wenatchee and Ephrata from 1935 until her death in 1977, died of colorectal cancer.

One of the girls in my high school graduating class died of papillary thyroid cancer, the same type of thyroid cancer that I had, before she was forty. She was raised east of where I lived, near the Moses Coulee area. There were only thirty-one in my graduating class. One died of leukemia in his early forties, and one died of pancreatic cancer just before he turned fifty. With my diagnosis, that amounted to thirteen of our class with cancer before age fifty, with 10 percent already deceased.

I'm a health professional in the Tri-Cities. I look at people, see a scar on a person's neck, and say, "You've had your thyroid out!" They are always surprised that I know that. I tell them I have had my thyroid out as well. You wouldn't believe how many people have had thyroidectomies in the Tri-Cities area.

One of the people who came into my place of employment had lived in Finley, near the Tri-Cities. He told me about pink snow in the late 1940s or early 1950s. I'm thinking maybe it was radioiodine-stained snow. He also told me about a lot of people with thyroid cancer. He had had a thyroidectomy as well.

I went to college with two people who were from the Basin City area, near the Tri-Cities, and I heard about all the stillborn children there with birth defects.

A man in my church who was raised in the Tri-Cities had a daughter who developed leukemia when she was less than two years old. The treatment stunted her growth and damaged her lungs, and she ended up needing to have a lung transplant. When she was in her twenties, she died from organ rejection of the transplanted lungs.

My dad lived to age eighty-eight, and my grandfathers to eighty and ninety-eight, so I should have had a chance at a long life, but with a big traumatic exposure like this, who knows how much it will shorten my life?

It's a full-time, high-maintenance job just making sure things in the body are still working as best they can under the circumstances.

Hanford's operations have hugely and negatively impacted my life. When I was going through the cancer treatment, my daughter was in high school. She cried because she thought her dad might be dead in a couple of years from the cancer. She wouldn't even hug me, she was so scared of losing me. And right about then, I really needed hugs.

The downwinders were never told the truth. The government lied to us and covered up the radiation releases from Hanford. They put us in harm's way and failed to protect us.

During World War II, production of nuclear weapons for purposes of national defense was understandable, but once the war was over, I'm not so sure continued manufacture of nuclear weapons to the detriment of the population was justified. Plutonium production at Hanford went on for decades after the war ended, throughout the Cold War era. That's the period during which I was exposed.

The downwinders are living with a lot of disabling health issues likely caused by our exposure. There aren't any resources out there to help us. The litigation didn't deal with all the other cancers, the effects of harsh cancer treatments, and other serious health issues the downwinders have suffered.

We need to put our stories out there publicly so that people know the full range of health problems Hanford may have caused.

Someone must be held accountable. The downwinders deserve, at a minimum, an apology from the government.

6

In the winter of 1961, on a night referred to as the "Night of the Little Demons,"[1] on farms downwind from Hanford's chemical separations facilities and production reactors, ewes suffered through unusually long and painful labors, and hundreds of new lambs were born dead, stunted, or severely deformed. Many were born without eyes or mouths, with crooked legs, or legless. Others had two sets of sex organs, while some had none. Many of the dead lambs had fused bones, and farmers had to reach inside the ewes' wombs to break the fetuses into pieces and pull them out.[2] Lambing ewes died as well, their wool falling off in clumps. Not only was the sight of so much suffering and death among their animals extremely hard on the Hanford sheep farmers and their families, but the loss of so many animals led to major financial hardship for these families, just as the deaths of lambs and ewes downwind of the NTS in 1953 had caused near or complete financial ruin to sheep farmers in Cedar City, Utah.

"Those Sons of Bitches at Hanford Killed My Sheep"

At the time of the sheep and lamb deaths in 1961, Hanford sheep farmers had no visible or tangible evidence that any radiation had been released off-site from the Hanford facility, but farmer Nels Allison was convinced that the nuclear facility, just across the Columbia River from his farm, was the cause of the deaths of his lambs and ewes. He had been raising sheep since 1957 and had never seen anything like this. "I told [a neighbor] that they was lettin' some juice out over there [at Hanford]," Allison said. "But [the neighbor] didn't want to rock the boat. In them days, we just took our lumps."[3] Allison lamented, "Those sons of bitches at Hanford killed my sheep and they almost made me lose my farm."[4]

In the early 1960s, a series of undisclosed reactor and plutonium processing plant accidents occurred at Hanford, releasing radioactive iodine and other radioactive isotopes into the air and the Columbia River. One of those accidents, at Hanford's PUREX chemical processing facility, occurred when the Hanford-area ewes would have been pregnant.[5] A United Nations report reveals that malformations in animals can be induced by exposures of less than five rems[6] of radiation.[7]

Animals experience relatively short latency periods, referring to the time following exposure to low-dose ionizing radiation or other toxins before the health effects of the exposure are seen.[8] For this reason, throughout history, animals, like the classic canary in the coal mine, have been used to warn of the presence of toxins. Sheep and other animals downwind of the NTS and Hanford showed evidence of the damaging effects of radiation exposure long before those effects began to show up in humans who had been exposed. The tragedies of the sheep and lamb deaths in Cedar City in 1953 and the farmlands downwind from Hanford in 1961 were early warnings of what was to come for the downwinders.

Hanford-area sheep farmers would not have known of the secret experimental animal farm hidden away in the interior of the Hanford site. They would have had no knowledge of the radiation experiments on pregnant ewes and their newborn lambs conducted there by Dr. Leo K. Bustad beginning in the spring of 1950.

Brenda Weaver remembers that night when the lambs were born without eyes, with deformed and fused legs, and other deformities, many of them on her family's farm. Her daughter, Jamie, was born four years later with a major birth defect.

Here are their stories.

Plaintiff 13: Brenda Weaver

I was born in Myrtle Creek, Oregon, in 1947. My dad was a World War II veteran. He had been a logger before he went into the military, but when he got discharged, he decided to give up logging because he wanted to become a farmer. At the time, under a new government program, farm units within the Columbia Basin were being offered in a lottery, with preference given to vets.

Water for irrigation of the arid desert lands of the Columbia Basin was pumped from the Grand Coulee Dam on the Columbia River. The government held a drawing, vets put their names in, and those who had their names drawn were able to buy farmland at very reasonable prices. My dad's name was drawn, so in 1953, when I was six, we moved to Moses Lake to begin farming. Winners in the drawing had to put up a dwelling, live there, and develop the land.

Hanford is located about eighty-nine miles to the south of Moses Lake, and the wind frequently blew in from the direction of Hanford. Violent windstorms could sometimes cause complete blackouts from the sand and grit, making it really hard to see or breathe. The tumbleweeds that blasted around in all directions during these storms were really painful when they hit.

By the time I was nine, in 1956, the farm in Moses Lake had gotten to be too small for our growing family, so we bought a larger plot of land in Eltopia, also within the Columbia Basin farmland program. Veterans' families from Utah, Idaho, Oregon, and other parts of the United States came to the Eltopia area to take advantage of this opportunity.

I attended Eltopia Grade School. Tom Bailie was in my class. Eltopia Grade School housed sixth, seventh, and eighth grades, with fewer than a hundred kids in the whole school. Up the hill, there was another schoolhouse, housing first through fifth grades, where we went to get hot lunch every day.

In Eltopia, as in Moses Lake, all of us kids played a lot in the dirt, and we ate fruit and vegetables grown locally or on our farm. We drank milk straight from our cows. I would imagine our food and milk were contaminated by airborne Hanford radiation. I know too that our fresh milk contained a much higher level of Hanford radioactive iodine than milk processed first at a dairy.

We swam in the Columbia, went water-skiing, and did a lot of fishing, so I think we got exposure from the river as well. One of my favorite places to go water-skiing on the Columbia was below Ringold, just across from Hanford. When I would fall into a warm spot in the river, it was wonderful because the river water was otherwise very cold. I had no idea those warm spots were caused by radiation discharged straight into the river from Hanford's plutonium production reactors.

A local doctor diagnosed me with hypothyroidism when I was only eleven years old and gave me a prescription for synthetic thyroid hormone. I have to take it for the rest of my life or I am in danger of lapsing into a myxedema [low thyroid] coma. There were tons of kids in the area back then with thyroid problems. I firmly believe that the government informed area doctors that they should check everyone for thyroid issues. I very much doubt that our doctors were ever told the reason for this. The government clearly knew the area's farming communities were being exposed to the constant barrage of radioactive iodine being released from Hanford as a by-product of plutonium production. Meanwhile, none of us had a clue that anything untoward was going on.

I also had a lot of respiratory problems when I was a kid, including repeated bouts of bronchitis. It got so bad that I was hospitalized with pneumonia when I was only eleven. My younger brother had really bad bloody noses all the time, and once, when he was only about four or five years old, his eyes began to bleed. My mom took him to a doctor in Spokane, but they couldn't figure out what was causing such a bizarre and worrisome medical problem.

When I was fourteen years old, I felt excruciating pain in my abdomen all of a sudden, so we went to the hospital, where we learned I had suffered a ruptured ovary. They rushed me into surgery. Fortunately, I still had one functioning ovary left. I have heard that whenever a thyroid or other organ was surgically removed or a biopsy was performed on someone from the Hanford area, the government preserved the specimen, and sent it to Spokane to some kind of Hanford-related tissue repository. I would bet my ruptured ovary was sent there to join the other biological samples.

My dad and Tom Bailie's dad were friends, and our farms were right next to one another. Most days, either Tom's dad would be at our house, or my dad would be at his. Both our families had sheep. Unbeknownst to the public, in around 1960 or 1961, there must have been a big release of radiation from Hanford. At the time, quite a few of our ewes were pregnant. A short time later, on the night they call the "Night of the Little Demons," our lambs were born without eyes, with feet missing, without mouths, some with legs grown together; baby lambs with all kinds of horrible deformities. They were monstrous in appearance. That's why they called them "little demons."

Only a few years later, while I was still living in the Hanford area, my first child, Jamie, was born without eyes. It is very telling that the lambs had been born without eyes, and then my daughter was born with the same birth defect. I believe that either my own exposure to Hanford radiation or my daughter's exposure in the womb, or a combination of these, caused her to be born without eyes. There was another child born blind in the Tri-Cities around this time as well.

I remember that shortly after the deformed lambs were born, men came out from Hanford wearing protective gear and carrying Geiger counters. They collected the dead animals and probably the deformed lambs. When the Hanford people came out like that, as they often did, we all thought they were just making sure we were safe. We figured if we weren't safe, they would have told us. It was the government's job to protect us, after all.

After Tom and I graduated from the eighth grade, we went on the bus every day to and from Pasco for high school. We graduated together from Pasco High School. When our graduating class held our fiftieth reunion recently, we learned that quite a few members of our eighth grade class have been diagnosed with cancer, and some have died. I'm one of those with cancer.

There were four kids in our family. I had an older sister and brother. My older sister died of ovarian cancer six years ago. My older brother passed away this past June of heart failure.

I lived in Eltopia until I got married. I married young and had two children in my twenties. My daughter, Jamie, is the oldest. Because Jamie was born without eyes, when I was pregnant with our son, Chad, two years after Jamie's birth, I was extremely frightened that he was going to be born blind like Jamie. My blood pressure climbed perilously during labor because I was so worried. The first thing I wanted to know after childbirth, when the nurse came in, was whether or not the baby had eyes. That nurse went backing right out of the room! She probably thought I was crazy to ask such a question. Turns out, my son, Chad, was beautiful, very healthy. I didn't have any more kids after that because I was afraid I would have another sightless child.

Jamie started kindergarten early, since she was so smart. In Spokane, where we lived at the time, there was an elementary school with a special classroom for blind students, so she went there for grade school. Eventually, she was mainstreamed into a normal classroom. Her dad and I fought very hard for her all through her school years to make sure she had the same opportunities as the other kids in her school.

Jamie was named national poster child for March of Dimes, 1974–1975. March of Dimes was a national organization dedicated to helping people with disabilities. Jamie was only seven years old at the time and very precocious. She was an animated and happy child. A press secretary and escort from March of Dimes always traveled with us. Onstage, he could get Jamie to react to things he said, and her responses were often quite funny. Audiences loved her.

During Jamie's tenure as national poster child, Nixon was president, then Ford. We actually went to the White House and met Betty Ford. When Nixon's continued tenure as president became questionable, our March of Dimes escort would get Jamie onstage, and one of the things he would ask her was whether she wanted to meet the Nixons. She would reply with an emphatic "No!" That always caused a lot of laughter from the audience.

When Nixon resigned, there was one day without a president before Ford took office. Carl Albert, Speaker of the House, was the most important man in the country that day, and that was the very day we were scheduled to meet him! We waited for some time outside his office while Jamie played with her dolls. We finally got to go in and found Secret Service men lining the office walls! Carl Albert was a short, freckled guy, and he kind of leaned back on his desk, wearing this big ring, I remember. I thought maybe I should kiss his ring! It was exciting to meet him on that historic day.

During the year she was national poster child, Jamie and I traveled to forty-some states. We met with three presidents, sixty-eight senators, TV stars, and famous athletes. We returned to Washington, DC, several years later for a reunion of poster children, and that is when Jamie met the Reagans. My husband and I didn't meet the Reagans, as Jamie could only take one person with her, and we let her brother, Chad, accompany her instead. We have a picture of Chad shaking hands with President Reagan.

Brenda and Jamie Weaver. Photo reprinted with permission of the *Oregonian*.

Because she gave so many public presentations as March of Dimes poster child, Jamie became a very good speaker. She wasn't afraid of an audience. She gave talks in our church as well. She still gets asked to attend different conventions and do presentations, and she has gone overseas to give papers as well.

Both my children chose to attend Brigham Young University in Utah. It seemed to my husband and me like a good, safe place to send our kids to college. Jamie, who was older, started there before Chad. In 1987, when he was just nineteen, Chad was killed by a drunk driver. He wasn't even a year out of high school. He had his whole life ahead of him.

It took us years to even begin to cope with Chad's loss. Jamie went to a therapist, and I think that helped her. In spite of losing her brother, Jamie managed to continue her studies at Brigham Young, completing a degree in pedagogy. She then transferred to Texas Christian University, where she got her master's degree. She went on to earn a doctorate at the University of Oregon, focusing on Italian composers and early classical music. She was then awarded a Rhodes Scholarship to study music in Bologna, Italy. In the beginning, I went to Italy to help her find her way around. People there didn't seem to expect blind people to be able to function independently. Adapting to a foreign culture as a blind person was rough for her at first.

Jamie is now a professor of music history. She is truly a fabulous person. She has prosthetic eyes she can use to appear more "normal." The only physical disability she has is that her eyes are missing. My husband, Jim, and I have always had high expectations of Jamie in spite of her blindness. We hoped that by expecting so much of her, she would grow up to be an independent person. She has met and exceeded our expectations.

With the technology of today, Jamie's life has changed for the better. Back in her high school years, very little was available in the way of assistive technology. We bought her an electric typewriter back then that was self-correcting. We were thrilled to have it, and it allowed her to type all her papers herself. Her teachers didn't take off points for mistypes. She was in the writing club, and she enjoyed writing a lot. She really has skill.

Nowadays, both her computer and smartphone talk to her. She has a new special remote for her TV that operates with voice commands. The TV narrates what is happening on the screen, and she can order Netflix by voice. She also has a FitBit on her wrist that she wears to help her keep track of her exercise routine.

Jamie's blindness isn't the only way Hanford's radiation has harmed our family. Both Jamie and I have hypothyroidism. I am monitored for thyroid nodules that might be cancerous, and any suspicious nodules are biopsied. I found out in October 2013 that I also have breast cancer. The cancer surgery was in January 2014. Following the surgery, I went through six weeks of chemo and an additional six weeks of radiation. I developed lymphedema [obstruction in the lymph system, causing swelling] as a result of the cancer surgery and radiation, and I go every week to have treatment. They were able to get all the cancer. I was very lucky it had not spread into my lymph nodes, although they took out two lymph nodes as a precaution. I am being monitored to make sure the cancer does not return.

I have other health issues that I feel are related to childhood radiation exposure. I was having some rather serious pain, so they checked my kidney, liver, and gallbladder. They can't decide if I have a polyp on my gall bladder or a kidney stone. I also have truly disabling fibromyalgia and osteoarthritis that I have had for decades. I was diagnosed with fibromyalgia twenty-five years ago. I have had the osteoarthritis since I was in my thirties.

The fibromyalgia moves around my body and causes acute pain. I can't take any of the medications normally used for treatment of fibromyalgia, as I react negatively to most drugs. I can't even take aspirin. I am allergic or reactive to a lot of things. I've had surgery on my sinuses three times. I have scarring on my face to prove it! Jamie has serious sinus issues as well. I've had polyps taken out of my nose. I also have asthma, which may have begun with the respiratory problems I had as a kid.

I have many digestive problems. I feel better when I don't eat gluten or lactose. I have to take antacids every day to help combat my stomach issues. I think some of the digestive problems go far beyond lactose or gluten issues.

I have had skin cancers removed several times. My brother and my mom also had skin cancer, and each lost part of the nose when the cancer was removed.

My husband, Jim, spent several years at Columbia Basin College in the Tri-Cities. He now suffers from Parkinson's. I don't know if the disease has any relation to his time near Hanford, but I do know that Parkinson's is considered exposure related.

After all I went through, with my daughter's blindness, my son's death by a drunk driver, and all these health issues, I had to decide whether I was going to live or die. I made the decision to live. It took a long time to get to that point. We were involved in a lengthy lawsuit over our son's death because the police were in a high-speed police chase with the drunk driver when the drunk driver, in his big pickup truck, plowed at eighty miles per hour into our son and his girlfriend. The police were defendants in the suit, as was the driver of the pickup, who at the time was on probation for a previous DUI. In the end, he was given just five years in prison. After Chad's death, I was involved with Mothers against Drunk Driving for something like sixteen years and did a lot of public speaking for the group.

I feel that Hanford downwinders are living leftovers of the nuclear age. We are the skeletons, the graveyard of what is left. As a child, I lived in what is known as the "death mile" in Eltopia, where there were one or more family members with cancer in almost every house along the one-mile stretch of road in our neighborhood. Those of us, like Tom and me, who lived there as kids, were highly exposed to Hanford radiation. Children are especially vulnerable to the effects of radiation. There are also children who lived in Walla Walla during the Green Run in 1949 and in Moses Lake, Spokane, and places around the Tri-Cities who were significantly exposed. The wind blew in all directions, after all, carrying with it radiation from Hanford.

People need to know that the radiation releases from Hanford were substantial and that real lives were shattered and destroyed by the downwind health effects of Hanford operations. The government covered up the truth for decades. I believe that all the land that was sold to veterans and their families was offered to us so we could become guinea pigs. I think the reason they brought so many people, primarily Mormons, up from Utah is because the Mormons are clean-living, and they therefore are better specimens in which to study the effects of exposure. I am convinced the government secretly watched and monitored us, all the while exposing us chronically to significant levels of radiation.

They would send those guys out from Hanford in white radiation protection suits with Geiger counters. They would test the ground and the

Hanford workers monitor farmlands for radiation, Atomic Energy Commission.
Photo courtesy of Library of Congress.

crops, pick up the dead and deformed animals, and take everything back
with them for even more testing. They knew what was going on; they knew
whole families were being irradiated out there on the farms. My dad was a
vet, and when we got the opportunity to move to Moses Lake and then to
Eltopia to farm with the Columbia Basin farmland program, we thought
this was the government being good to the vets and honoring them. In

reality, it was the government using the vets and their families as guinea pigs. It was a way to populate the area just across the Columbia River from Hanford with human exposure study subjects. I'm convinced of that.

Sometime after Chad was killed, I attended a conference sponsored by the DOE. The DOE invited people from downwind populations at different Manhattan Project sites, including Oak Ridge, Rocky Flats, Hanford, the Nevada Test Site, military people who had witnessed atomic tests, Navajo uranium miners, and several people who had been the subject of individual human radiation experiments. We were all there to share our stories. At these meetings, government officials collected all our information, and we honestly believed they were going to help us.

At another DOE meeting, a scientist who worked with the DOE and had conducted research on the topic told us that he concluded that the reason there were elevated levels of multiple sclerosis in areas around Hanford was people's exposure to nuclear fallout from Hanford, Nevada Test Site atomic testing, and global testing. He said he had proof. The DOE fired him shortly after he made these assertions. Fortunately, he later got a professorship at Berkeley. He spoke to us by satellite at one of our meetings, and we could ask him questions. What he shared with us confirmed my belief that all these health issues were from Hanford.

I'm truly angry because my daughter will spend her whole life blind. Unless they come up with a means to replace a whole eye, there is no hope of sight for her. I don't think she has any functional optic nerve to tap into even if they were to invent a prosthetic eye. I'm angry because of what she has had to go through and will have to go through the rest of her life. She has made a wonderful life for herself, but it has been a lot of work. And there are still many obstacles she has to overcome every day.

People say, "Oh, man, she's amazing," or "Oh, this" and "Oh, that." Yes, she is amazing, but they don't know what she has to go through just to get up and go to work every day. They don't know what she has to go through just to get things done. As a professor, it takes her much longer to do the things a sighted person can do. She does it all on the computer, grading papers and all that, but it takes her twice as long as a person who is sighted.

Opportunities in her life have been limited. She's not married. It seems like women will marry blind men, but men don't want to marry blind women. Isn't that strange? Women will take care of people with disabilities, but men don't want to take care of people with disabilities so much. She hasn't opted to adopt, as she believes it isn't fair to raise a child without a father.

Jamie has faced a lot of discrimination throughout her life. People who haven't known people with disabilities often don't know how to interact with them. People who live with adversity in their own lives learn how to have empathy and understanding, but if you have never faced adversity, you don't know how to interact with people who are different in this way.

In high school, she got her cane taken away, and she got felt up by a group of boys. She has been bullied in other ways too. It took a while to find the boys who did it, but she finally recognized their voices and turned them in. They didn't have self-confidence, so they took it out on people who were vulnerable, like her.

It's really not easy being blind. It's doable for her but not easy. My heart breaks for her to be able to see certain things. She used to say when she was little, "Mom, how is it you say the clouds look like cotton balls, but we can fly through them on the airplane?" Once she asked, "What is energy?" and "How come you say we can see the bottom of the stream and the rocks that are down there, but the water is blue? How can you see through blue?" I discovered she could actually see stars because, when you are dizzy, you see them with your brain, not your eyes. I told her that is what stars look like, those little points of light you see when you are dizzy.

When she is out with her guide dog, depth perception can be a problem, especially crossing streets. I'm also very concerned about her in the future because they are making all these electrical cars, and their motors are virtually silent. You have to be able to hear a car to know it is there when you are blind.

We are a good family, and we gave her a lot of opportunities and built her confidence. We taught her to snow ski and water-ski. In fact, my husband, Jim, almost drowned teaching her to water-ski! When we went to the zoo, we asked to touch some of the animals. When we went to the circus, we got permission to go back and touch the elephant. We tried to give her a whole lot of hands-on experience, chances to touch and explore things, to help her learn about them.

In the United Kingdom, where Jamie had a talk to present at the University of East Anglia, we stayed for three weeks and went on a number of tours. One tour was of Windsor Castle, and we were allowed to go in

Brenda Weaver. Photo by author.

and touch busts of people and statues. They also let her in past the ropes to touch the bed that Napoleon and his wife slept in when they stayed in the castle as guests. It's very rare to be allowed past those ropes! I appreciated the open-mindedness of the staff guarding the castle exhibits. People weren't always that open-minded. It depended on so many things.

Once, on Lake Roosevelt, she got to steer a Jet Ski, a two-person, where there was nothing for her to run into. She said she had never felt so free! She also took horseback riding lessons with another lady alongside on another horse.

But in spite of all these opportunities, it doesn't take away from what we have to do to help her, even today—when she buys new clothes, I have to label all her clothes with braille labels so she can color match. For a man, it isn't so hard, as they only have one or two suits, or brown or black socks to match. Women, though, usually have a lot of shoes and outfits. So I have to label all her clothes after she buys them. I use a metal label in braille. The plastic labels melt in the dryer. I add a braille label in place of the original clothing label.

What has happened to the farming families and their children, downwind of Hanford, is a gross injustice. We were misled, and we were used. My daughter will spend her life as a sightless person because of it. I feel very badly for all the people who have suffered with the devastating health effects of these exposures. Our children and our grandchildren are all victims in this as well.

Rather than honoring veterans with agricultural land within the Columbia River Basin, the government farmland program downwind of Hanford has rendered us victims of the nuclear age, and many of us now suffer from cancers, other significant disease, and birth defects.

Our government covered up the radiation releases from Hanford for decades. In those meetings of exposure victims sponsored by the DOE, we all thought the government was going to help those of us harmed by the US nuclear weapons production and testing program. Instead, it seems like they are just waiting for us all to die.

Plaintiff 14: Jamie Weaver

I was born in Wenatchee, Washington, in 1965, although I was conceived in the Tri-Cities, the communities closest to the Hanford nuclear reservation. I am both the granddaughter and daughter of Hanford downwinders and am a downwinder myself.

I was born with an extremely rare birth defect called anophthalmia, a birth defect that causes sightlessness. This isn't the same as being born blind or visually impaired or with partially formed eyes. This means I was born with completely empty eye sockets—no eyes and no optic nerves.

When I was born, and the doctors and nurses realized I had no eyes, they had no idea what to do, so they whisked me away and didn't let my mother see me until the next day. There was some real sexist stuff going on in that my dad was told about my birth defect first. I guess they thought my dad, as a man, could take the news better than my mom, as a woman. Then, my grandparents on both sides were told, and finally, my maternal grandma told my mom. My parents were both very young, and I don't think anyone knew how to break the news to them that their first baby had a significant birth defect.

Anyway, both my mom and my grandma ended up crying and crying, and my mother kept saying she didn't know what to do to take care of a blind child. She said the only contact she had had with blindness was reading the Helen Keller story.

My mom said when the nurse finally brought me to her, I looked normal, as I was asleep, with eyelids shut. I had baby-fine red hair and perfect little fingers, hands, and toes. The lack of eyes was my only physical deformity. I was otherwise quite healthy.

I'm pretty convinced that I was born with anophthalmia because of my mother's exposure to Hanford radiation. She lived on the family farm in Eltopia, just across the Columbia River from Hanford's nine plutonium production reactors. They got significantly dosed there. She remembers men with Geiger counters wandering around all the time, taking samples of soil, cow and goat milk, vegetables, fruit, and other farm crops. She always thought that all this surveillance and testing meant the government was taking good care of everyone in the area.

She also remembers that her class at Eltopia Elementary School underwent a lot of medical testing at school. There was both a doctor and a nurse at the school for those exams.

Ezra Taft Benson, secretary of the Department of Agriculture under Dwight Eisenhower, encouraged Mormons to move to the Columbia Basin area to farm. A lot of the farm families were there as well because of the federal land grant program that, through a lottery, gave vets land in the area to farm. This was supposed to be a reward for their service to our country. So many of the people, like my mom and her parents, ended up getting exposed out there in the farmlands, it wasn't much of a reward!

My mom said that one night around 1961, there were farm animals born on their farm and on neighboring farms with major deformities—fused legs, deformed limbs, and some without eyes. This night was always referred to as the "Night of the Little Demons" because the deformed animals looked like little monsters. Sometime before these births, a large amount of radiation was released from Hanford. First, the farm animals were born without eyes, and then, not long thereafter, I was the one born without eyes! It's hard to ignore the possible connection between Hanford's radiation and this birth defect shared by baby animals and me.

The doctors at the hospital where I was born tried to figure out what had caused my anophthalmia. They told my mom that she must have had rubella [German measles] or received a rubella vaccination during pregnancy. She said no, she hadn't gotten any vaccinations or shots, and she hadn't had any of the spots characteristic of rubella or been sick at all during pregnancy.

Rubella seemed to be their only theory. They didn't offer my family any other explanation.

My parents have always been supportive and awesome with me. They labeled things for me with tactile materials, made sure I had a braillewriter and any other assistive devices that existed at the time, and enrolled me in a special class for children with visual impairment. They made sure I had every opportunity. One of those opportunities was with March of Dimes, an organization that works to prevent birth defects. When I was about five or six years old, I began working with them, first as the Washington State poster child and later as the organization's national poster child for 1974–1975.

As the state poster child, I often gave presentations with a biologist who worked with March of Dimes and was also a professor at a local university. He examined me in order to try to answer my parents' questions about what could have caused me to be born without eyes. He started asking us all these nuclear-related questions, like if I was conceived near a nuclear facility, where my parents had lived earlier, that sort of thing. He didn't say why he was asking, and no one knew yet about Hanford since the DOE didn't make their documents public until 1986, and this was around 1971.

My parents told him that the doctors in Wenatchee had thought my anophthalmia was related somehow to rubella. The biologist told us that rubella causes one of two things, either brain damage or blindness, but that rubella does not cause missing eyes. It's funny, he always seemed guarded in what he said, like he knew something, but he didn't want to upset us. One thing I like about March of Dimes is that they involve a lot of scientists in their work and take a very scientific approach to birth defects—trying to get to the cause in order to prevent these birth defects in the future.

Am I angry about what Hanford has done to me? There are days that are really hard, living sightless in a sighted world, and I guess the government caused that. And there isn't really anything that can be done to restore my sight—I don't have an optic nerve to tap into if there were a functional prosthetic eye developed that might work for me.

Sure, I have some pretty awesome adaptive technology now, particularly with apps on the iPhone, but honestly, most things take longer for me than for most people. I also have a guide dog, and that really helps me, but honestly, being sightless is not easy. I did earn my PhD and now teach at Stephen F. Austin State University in Texas, and I credit a lot of my independence to my parents and their high expectations of me.

I feel that the government willfully and intentionally used its people as test subjects. I remember being told by someone who grew up with my mom that if I went to Los Alamos and looked at the nuclear files they keep there, I would find a file with my name on it. That is so creepy to think about.

Just after the DOE admitted in 1986 that Hanford had released so much radiation onto communities downwind, my mom went to a meeting sponsored by the DOE where representatives for the downwinders gave personal testimony. At the conclusion of several days of testimony, DOE representatives said, "We have listened to your stories; now we want to know what you want."

The lead downwinder representative said that the downwinders wanted the DOE to promise not to perform this kind of testing on people again. The DOE people said they couldn't promise that. I realize that government people cannot say "never" and "always." But what bothers me is that we may not have learned from this experience. Also, I think that the downwinders thought the government was going to help them, and that is why they traveled all that way to give their personal stories. No help resulted from that meeting.

I don't think the story of the downwinders is finished. I think that their

Jamie Weaver in the role of Helen Keller in a Spokane theater production. Photo courtesy of Weaver family.

Jamie with her guide dog. Photo courtesy of Weaver family.

children and maybe even their grandchildren will have health issues down the road—health issues that were probably caused by their parents' and grandparents' exposure to Hanford radiation.

The most important thing to me is that society should learn from what has happened to all of us. History is only worth studying if we learn from it and don't make the same mistakes again. It seems that with the Hanford downwinder situation, we are not doing that. Neither the government nor anyone else has studied the range of diseases and birth defects in all of us who have been exposed to Hanford radiation. There was only the HTDS, and that was inconclusive.

I think that the government doesn't want to study the downwinders because they are afraid of what they may find.

Brenda's and Jamie's stories are heartbreaking. The Weavers have made the best of an incredibly challenging situation, raising Jamie to be an accomplished and independent woman. Yet beneath this strength is a wealth of pain and a certainty that these monumental challenges that life has presented them very likely arose from the near obsession of the United States during the Cold War era with building a greater stockpile of nuclear weapons than the stockpile amassed by the Soviet Union—a quantity of nuclear weapons that could destroy the world many times over at a moment's notice. As this nuclear obsession continued unabated, people like Brenda and Jamie Weaver became the innocent bystanders sacrificed in the process.

Testing Hanford's Irradiated Children

Of special interest to Hanford's environmental monitoring program were the small farming towns of Ringold, Eltopia, and Mesa. Ringold was one of the few farming regions that made use of Columbia River water drawn downstream from the Hanford reactors for irrigation purposes.[9] Ringold included about five hundred acres on a flat area between the Columbia River on the west and high bluffs on the east. To the north was the Hanford project boundary, and to the south, the river met the bluffs. The area was about thirteen miles downstream from the nearest production reactor and thirteen miles east of Hanford's chemical separations plants.

Throughout the early 1960s, Hanford scientists regularly collected samples of produce grown on Ringold, Eltopia, and Mesa farms and distributed dietary surveys to the residents of Ringold without informing them that the information would be used to estimate their cumulative radiation exposure from Hanford's airborne and river-borne radiation. The surveys tracked length of residence in the area; sources of water, milk, and fresh produce and quantity of each ingested per day; and intake of fresh seafood and fish caught in the Columbia along with the specific location of the catch.[10] Ringold produce was radio-assayed, and the results were used to develop estimates of the radiation exposure of hypothetical Ringold residents, applying assumed dietary habits.[11]

Area residents were likely told that the dietary surveys and collection of homegrown produce were being carried out to ensure that all was well in Hanford's environs. People in the area were generally aware of Hanford's role in the wartime Manhattan Project, and they knew that the facility still operated in some Cold War capacity. Hanford officials would have assumed that any concern in the region over undisclosed radiation releases from the facility would be allayed by the monitoring carried out so publicly throughout the Tri-Cities and Hanford-area farm communities.

Meanwhile, Hanford continued to secretly release radiation to the air and the waters of the Columbia. People in the area could have absorbed radionuclides in multiple ways, including from inhalation, from ingesting Columbia River water downstream from the reactors, from drinking milk containing I-131 or radioactive zinc,[12] from consuming meat or produce raised on land irrigated with radioactive Columbia River water, or from consuming contaminated fish and waterfowl from the river. The absorbed radionuclides for which exposure dose was calculated on dietary forms distributed to Ringold residents included radioactive isotopes of sodium, potassium, zinc, and cesium.[13]

In 1962, Hanford added the use of a whole body counter (WBC) to its assessment of exposures in off-site downwind communities. The first Hanford WBC was constructed in 1959 within a seven-ton iron room measuring about ten feet square with ten-inch-thick steel walls, ceiling, and floor.[14] A mobile Hanford WBC unit was built in 1963. In spite of claims that use of the WBC "avoids any of the assumptions necessary in the usual exposure estimate,"[15] the WBC measured only gamma rays, not the beta rays also emitted by I-131,[16] and was unable to detect whether the individuals tested had been exposed in the past to the many short-lived radionuclides released from Hanford. These short-lived radionuclides could have done damage in a body, but, by the time of testing, the radioactivity would have decayed. Short-lived radionuclides are often more dangerous than long-lived radionuclides, as they can deliver a heavy dose of radiation in a short time.[17]

Ringold families were invited to come to the Hanford facility to be examined in the WBC. This invitation was issued in 1962, well after the years of highest radiation releases from Hanford. Twelve Ringold residents, including several children, agreed to be examined. Gamma rays emitted by zinc-65, cesium-137, and I-131 in the thyroid were measured. In spite of the limitations of testing I-131 levels in the WBC, two children tested had substantial levels of I-131 in their thyroids, and their parents were advised that the levels, although elevated, were within "permissible levels."[18] Hanford officials failed to mention that these permissible levels applied to adult workers in jobs involving radiation exposure and that they were far higher than the "permissible levels" in growing children.[19] The families went away from the testing session reassured. Hanford officials then published a paper claiming that these results were "gratifying."[20]

The official follow-up report on the testing did not identify Hanford's airborne releases as the source of the I-131 detected in the Ringold residents' thyroids. Instead, fallout from the NTS and global testing was blamed: "Fallout from weapons testing was relatively high in this area at the time of examination which accounts for the occurrence of I-131 in the thyroid glands."[21] The high levels of I-131 released from late 1944 through the late 1950s from Hanford were no longer present in Ringold residents' thyroids at the time of testing in the WBC due to the short eight-day half-life of I-131. It seems more likely that the I-131 present in the Ringold residents' thyroids represented a combination of Hanford I-131 and I-131 from global fallout. It is not known how much higher I-131 thyroid levels would have been had they been tested during the years of the highest I-131 releases from Hanford or with a direct thyroid scan rather than the WBC.

In fall 1962, Hanford conducted an additional study of I-131 levels in the thyroids of thirty-eight area children aged two and a half to seventeen years. The studies were prompted by the easily measured levels of I-131 in local milk.[22] The results revealed that about 20 percent of the I-131 in local milk had been taken up by the children's thyroids.[23]

In 1968, Hanford initiated a study of elementary school children, ages six through twelve, and a smaller number of children ages thirteen to fourteen throughout the Tri-Cities. The study, "Influence of Diet on

Radioactivity in People," involved a classroom presentation to the children on radiation and the completion of a dietary form for a week, followed, with parental permission, by a scan of each child in the mobile WBC parked on the school grounds, where the body burden of internal radiation was measured.[24] It is not known whether these tests caused the parents of the Tri-Cities to be concerned about possible radiation exposure in their children. The children studied within the Hanford area represented only a small percentage of the children within Hanford's vast downwind region who had absorbed a range of radionuclides released from Hanford.

Tom Bailie and his sister Mary grew up on a farm located between Ringold and Eltopia. Tom remembers deformed animals born on the family farm and how the "people from Hanford" would collect "weird stuff" such as dead chickens, vegetables from the family garden, and water samples. Tom and Mary both suffer from serious health issues that they believe were caused by their childhood exposures to Hanford's radiation releases.

Here are their stories.

Plaintiff 15: Tom Bailie

I was born March 10, 1947, at Lady of Lourdes Hospital in Pasco, Washington. We moved into our farmhouse near the town of Eltopia when I was four years old. The farm was ten miles east of the Columbia River and four miles west of the small town of Mesa, across the Columbia River and directly downwind from Hanford's eight plutonium production reactors. Prior to age four, I lived on a farm even closer to Hanford.

My mother had a miscarriage in 1945 before getting pregnant with me. A lot of women in the area had miscarriages that year. I was born in 1947 with minor deformities—malformed nails, sunken chest, underdeveloped lungs, and slightly bent limbs.

I was a pale, sickly kid and was in and out of hospitals throughout childhood. I had skin sores that wouldn't heal, staph infections, and bloody noses all the time. I had asthma, allergies, and chronic diarrhea. At age five, I nearly died of a mysterious paralysis. I recovered only after spending two weeks in an iron lung, allowing me to breathe. What I had was diagnosed as polio, but I didn't have the atrophy of polio. I'm convinced I had something other than polio. The paralysis lasted about five months.

Tom Bailie's father on the farm. Photo courtesy of Bailie family.

When I could breathe on my own, they first put me in a wheelchair and then fitted me with leg braces like Forrest Gump wore. I started the first grade with those braces on. I wanted so badly to be normal and to play with my friends that I pushed myself to grow stronger, and eventually, I was able to walk without the braces.

I was married and fathered a daughter at age seventeen, but a sperm test at age eighteen revealed I had since become sterile.

My mother gave all of us Bailie kids some kind of pill every day. It was a thyroid pill or maybe an iodine tablet. My mother was one of Dr. Oppen-heimer's secretaries in Berkeley, so she may have known something about the airborne radiation being released from Hanford. None of us have thy-roid problems now. We seem to have escaped thyroid damage from Han-ford's radioiodine thanks to that daily pill. We have a lot of cancers and other health issues but no thyroid problems. All the other kids who lived around us have thyroid disease or thyroid cancer.

I remember some kind of testing in grade school, looking at our necks. I am sure it was some kind of thyroid test. We always knew something was wrong. Some of my earliest memories are of the deformed animals born on our farm. When the deformed animals were born, my mother used to say, "Must be that damned Hanford again." In the 1950s, a gray Studebaker would sometimes pull into our driveway, and my parents would say, "It's those people from Richland again; the Hanford people are here." Mother made homemade pie for Dr. Norwood, Hanford's medical director, who always rode in the back of the car and brought Mother fresh-cut roses. One time he brought me red cowboy boots and asked what I had eaten that day.

Now I realize that Dr. Norwood never ate the pie my mother made! I always ate it! I remember how the people from Hanford always asked us for weird stuff—dead chickens, vegetables from the garden, water samples. They asked how much milk we drank and all that.

Back in 1956, my father owned fourteen thousand acres of land. The government awarded part of that land by lottery to World War II veterans so they could start a life and have a family there. I don't know whether they ever paid my dad anything for his land.

The lottery was part of the Columbia Basin land grant program. Families came to be part of the program from all over the nation, from all walks of life, from different backgrounds. I don't think any of them were farmers. All they had was the desire to prove themselves on eighty acres, raise their children there, and live the American dream, owning a piece of land and farming it.

The Farmers Home Administration required all of us to raise cows and establish small dairies. We called them the "AEC dairies" because every week, the AEC came in and took milk samples, tested the milk, and reassured us all that it was safe to drink. It is those vets and their families, along with those of us who already lived and farmed in the area, who were directly in the path of Hanford's radioactive fallout and who have paid the price with our health and, often, our lives.

I first found out about Hanford's radiation discharges in 1986 after the declassification of nineteen thousand pages by the DOE Hanford site manager Mike Lawrence, responding to public demands to know whether anything harmful had ever been released from Hanford. I remember that Lawrence stood up there behind the podium in Richland with his hand on a pile of documents about three feet tall and said, "There are no observable health effects." We all thought, "What the hell was that all about?"

He said, "There are no observable health effects from Hanford's historic radioactive releases." We had no clue those documents would reveal decades of radiation to the air and the Columbia River and that we were immediately downwind.

Karen Dorn Steele did the first Hanford downwinder story in 1985, a year before those documents were released by the DOE. The story was called "Downwinders—Living with Fear." Not long before she wrote that story, I drove with Karen through my neighborhood along what has come to be known as the "death mile." As we drove past my neighbors' houses and trailer homes, I listed for her all the cancers that had afflicted each family. Nearly every family had at least one member with cancer.

All of a sudden, while I was driving Karen around, it dawned on me—I had been in an iron lung. I have always had nosebleeds, chronic diarrhea, and asthma. I'm sterile. I've been sick all my life. I've always thought it was me. That's what's wrong! It's something in the environment here that's making us all sick. It's Hanford! It's that damned Hanford! I pulled the car over, and I said, "Jesus, Karen, that's what happened!" Karen asked me why I hadn't told her before. I told her it had never dawned on me before.

I know that the death mile tour motivated Karen's efforts to find out what was really going on at Hanford. I feel that the stories Karen wrote about the death mile and the downwinders were a big part of the reason the DOE, Richland was finally forced by public pressure to declassify thousands of Hanford records in 1986, records that finally revealed Hanford's decades of downwind and downriver radiation.

It wasn't just my neighbors and friends who were dying. My family was hit big-time. My father died of cancer; his brothers, Wayne, Ray, and Chet, and sister, Beulah, died of cancer; my father's mother died of cancer; his father died of cancer. We all lived in the same area and ate the same food and drank the same water. Virtually all my father's family died of cancer.

You can say, "Aha! Must be hereditary!" But it could also have been environmental because we all slaughtered the same chickens on the farm, slaughtered the same pigs, milked the same cows, grew the same vegetables; we all ate out of the dirt right there on the farm. My sister Mary has had cancer. I have had skin cancer along with my other illnesses. I have had leg surgery, treatment for undescended testicles, and birth defects—a

Tom Bailie on the family farm. Photo by author.

sunken chest. I was a crooked, twisted baby when I was born. So that is our family history—just full of cancer.

These health problems piss me off. They make me want to fight and live longer. I could never play sports like the other kids did. I could shoot better, though, since I had lots of time to practice. We had veterans come here to the farm, and I listened to all their war stories from World War II and the Korean War. My dad would go down to the railroad and pick up winos, and they would work the harvests. I sat there and smoked cigarettes and drank beer with the vets and winos when I was ten. We shot the sparrows out of the trees and the rats out of the elevator and hunted rabbits. I was never physically strong, but I got to be a damned good shot.

The downwinders were used by our government. They considered us a "low use segment of the population." We farmers didn't even know we were poor. We didn't even know we were abnormally sick. We were just good American citizens, many of whom went off as patriots to fight for our country in the Korean War and later, in the Vietnam War.

We are down at the bottom of the food chain here on these farms. Academically, our skill levels are not as high as the average American, but we feel we are worth more than they think we are. So I'd like people to know we contributed, we gave our lives, and as citizens we are every bit as good as everyone who worked at Hanford.

Plaintiff 16: Mary Bailie Reeve

I was born June 2, 1949, in Walla Walla, Washington, one of four children in a farming family. We lived in the farmlands that grew up after irrigation water was pumped in from the Grand Coulee Dam on the Columbia. Everything is identified out there by "blocks." We originally lived in Block 20, above Basin City, toward Othello. Eventually we moved down to the farm near Eltopia. Block 20 was even closer to Hanford than the Eltopia farm. Before the irrigation water, that was all just dry desert land, no trees, just blowing sand.

My brother Tom remembers the monitoring stations Hanford had set up in our area. These were little stations. I don't know exactly what they were monitoring.

When I was a child, I was pretty healthy. We Bailie kids all had our tonsils out, and my brother Tom was paralyzed for a while and spent time in

Tom, Mary, and their father. Photo courtesy of Bailie family.

an iron lung, but I was okay. In the third grade, I had a real problem with sinuses, and I had extreme headaches, but not much else.

Every day, Mom gave us kids something that protected our thyroids, and I think it was a pill. It might have been potassium iodide tablets. She might have known something since she had worked as a secretary with Dr. Oppenheimer in Berkeley.

Things went along okay until I was an adult. When I turned thirty-two, I had to undergo a hysterectomy. Two years later, I had colon issues. They found so many polyps in my colon that I had to have part of my colon surgically removed. Then, I was okay again until 2009, when they found a lump in my breast. I was diagnosed with breast cancer and had to have a

Mary Bailie Reeve. Photo by author.

double mastectomy. Luckily, I caught it early. I am pretty vigilant because of the problems I've had.

In 2014, they found more polyps in the only section of colon I still had. So I went to Virginia Mason and had my whole large colon removed. I had a J-pouch procedure and eventually had the remaining colon reconnected.

I had to have an external bag for several months. This was really the hardest surgery I have ever gone through, and I have had a few.

Nowadays, I am concerned about our water supply. I'm on a community well, and I am, at most, three miles from Hanford and the Columbia River. We are thirteen miles north of Pasco and seven miles south of where we used to live. I lived here most of my life. They always test the water and say it is fine, but I am worried that the contaminants from Hanford may have migrated under the river over to the well. So I only drink bottled water—I don't trust the well water.

Until you have breast cancer, you don't know that others have breast cancer—now that I have it, I have learned that so many women out here near Pasco have breast cancer. There were a lot of miscarriages out here as well. My mother had one, and I know a lot of other women who had them.

I feel that the government should study what happened to all of us. We have maps of all of our neighbors indicating where we grew up and who has died of cancers. There are an unbelievable number of people who have died from cancer out here.

At around the same time that the radiation body burdens of Hanford area children were being assessed in Hanford's WBC and their parents were being assured by Hanford and AEC officials that any findings of internalized radiation in their children were nothing to be concerned over, multiple epidemiological studies and a series of congressional hearings began to investigate whether there might be a connection between fallout exposures and leukemia and thyroid cancer downwind of the test site.

7

Concern about the health effects of fallout in communities to the north and east of the NTS led to the first congressional hearing on the topic in 1959.[1] There were early reports of health issues within these communities, but epidemiological studies of populations downwind of the NTS had not been carried out. Health concerns identified during the congressional hearing resulted in two federally funded studies of cancer prevalence in Utah. Neither of the studies attempted to calculate the fallout exposures of study subjects in order to determine whether these exposures were related to the cancers found, a calculation considered a minimal criterion for epidemiology.[2]

In the first of the two studies, Dr. Edward Weiss of the USPHS investigated the incidence of leukemia between 1950 and 1964 in Washington and Iron Counties, two significantly fallout-exposed counties in southwestern Utah, finding a 3.29-fold excess of leukemia among those under age nineteen and a 1.5-fold increased risk of leukemia in people of all ages.[3] The study counties were predominantly Mormon, a population that traditionally has a much lower cancer rate than the rest of the country, which made these results even more significant. Weiss did not attempt to correlate the AEC's fallout dispersion data with the location of the residences of children who had died in order to determine whether the excess leukemia deaths might be related to fallout from NTS atomic tests.[4] Based on a joint AEC–White

House decision, the study results, known to the government by 1964, were not made public. The AEC counseled the USPHS against further investigation of the leukemia deaths, hoping thereby to avoid public concern that might result in litigation, which could jeopardize continued testing at the NTS.[5] Although atomic testing at the NTS went underground in September 1963, there was still potential for public anxiety over atomic testing because the underground tests continued to vent airborne radiation. The results of the Weiss study were suppressed by the AEC until 1979,[6] when many AEC atomic testing records and minutes of AEC meetings were declassified for congressional hearings on the health effects of low-level radiation from atomic testing.[7]

In 1965, concerned that radioisotopes of iodine contained in NTS fallout had entered the milk supply downwind, the federal government commissioned a second study, led by Dr. Marvin Rallison of the University of Utah, to determine whether thyroid cancer had increased in children exposed to fallout in southwestern Utah and western Nevada. Approximately three thousand schoolchildren living in western Nevada and southwestern Utah and a control group of two thousand children living in southern Arizona were identified and examined for thyroid neoplasms over five years. The study is referred to as Phase 1 of the NTS fallout thyroid studies.

Rallison and his team did not obtain any information for the study on the source and amount of fresh cow's milk consumed by each child during the period of atomic testing, making it impossible to assign a thyroid radiation dose to any of the children. Critics have labeled this a gross oversight that is rare in epidemiology, comparable to studying lung cancer without asking about participants' smoking habits.[8] The Rallison study did not find an excess of thyroid cancer among the Nevada and Utah children studied. The AEC used the negative findings from this study to reassure the public that no cancers had resulted from atomic testing.[9] Rallison recommended follow-up study of the cohort due to the long latency period that typically follows exposure to low-dose ionizing radiation, suggesting that additional cases of thyroid cancer were likely to appear in the future among members of the exposed cohort.[10]

In 1979, in a study on fallout from atomic testing and childhood leukemias, a team of investigators led by epidemiologist Lynn Lyon from the University of Utah unknowingly replicated and expanded the findings of the suppressed 1965 USPHS Weiss study that had found excess leukemia in southwestern Utah.[11] Assigning children to either a high- or low-exposure group, Lyon's study calculated fallout exposure levels based on the residential history of the study participants rather than through the use of unreliable, incomplete, and possibly underestimated exposure data from the AEC.[12] Lyon and his team investigated leukemia deaths in Utah children who were under fifteen years of age between 1944 and 1975. The study found a 2.44-fold increase over the expected childhood leukemia death rate in seventeen high-fallout southern Utah counties between 1951 and 1958 and concluded that the "excessive leukemia deaths were due to ionizing radiation exposure."[13] Lyon and his colleagues became the focus of an intense effort by the federal government to disprove their findings, an effort that included reanalyzing the study data four times at a higher cost than the original study.[14]

Leukemia and the Atomic Vets

In 1976, Paul Cooper, a serviceman who had participated in military maneuvers following NTS shot Smoky, was diagnosed with acute myelocytic leukemia. Cooper was one of 250 troops who had gone within one hundred yards of ground zero minutes after detonation. Cooper's claim for service-related injury with the US Department of Veterans Affairs (VA) was repeatedly turned down. Cooper requested information from the VA about whether shot Smoky could have caused his leukemia.

The following year, Gordon Eliot White, a reporter for the *Deseret News* in Salt Lake City, did a story on Cooper, citing National Cancer Institute (NCI) statistics that demonstrated a small increase in cancers in Utah at the time of the atmospheric tests.[15] White's story hinted at a possible link between atomic testing and cancer, not just for Cooper but for others downwind of the test site. The story was picked up by the national media, resulting in growing public anxiety over the hazards of atomic fallout. Dr. Glynn Caldwell at the CDC told the *Deseret News* that he would track down veterans who had witnessed shot Smoky to study the incidence of leukemia in the group.[16]

In 1977, White wrote a story in the *Deseret News* referring to an unpublicized NCI survey from 1970 that revealed that leukemia deaths over a twenty-year period were nearly twice the state and national average in one county in Utah and significantly higher than average in the others. The survey unearthed by White was part of an NCI nationwide county-by-county investigation of deaths between 1950 and 1969.

In November 1978, White obtained classified reports revealing that the number of cases of leukemia in the southern Utah towns of Parowan, Paragonah, and Monticello far exceeded the number expected to occur naturally.[17] These reports were the result of a leukemia case reporting program in the 1960s in which Utah had participated. White described the unreported excess of leukemias in the *Deseret News*.[18]

That same month, Governor Scott M. Matheson of Utah met in Washington, DC, with President Carter and Joseph Califano, secretary of health, education, and welfare, to request help assessing the health effects of repeated exposure to low-level radiation downwind of the test site. Following the meeting, Califano read through AEC files on the atomic testing program, many of which had been classified by the AEC and its successor, the DOE. After sharing the contents of the files with Matheson, Califano agreed to make them public. Matheson was himself a downwinder who had witnessed atmospheric atomic tests in the early 1950s from his home in southern Utah. Family members who had watched those tests alongside him now suffered from cancer and other health issues. Matheson became a staunch advocate for the downwinders throughout his two terms as governor. In the mid-1980s, Matheson was diagnosed with multiple myeloma (bone-marrow cancer) believed to have been caused by his childhood fallout exposure. He succumbed to the cancer in 1990.

Congress Responds to Reports of Cancer Downwind

In 1978 and 1979, responding to growing numbers of reports of cancer in downwind communities and to epidemiological studies confirming the excess of disease, congressional subcommittee hearings on the health effects of atomic testing at the NTS were convened in Salt Lake City; Las Vegas; and Washington, DC.[19] Members of communities downwind of the NTS shared stories of cancer and other illnesses they believed to have been caused by fallout exposure. The hearings revealed that the AEC was aware of the health risks of atomic testing to communities downwind yet had failed to inform the public of the time and place of tests or of precautions that could have been taken to protect against fallout exposure and injury. It became clear that the atomic testing program had pushed ahead without concern for the welfare of downwind populations, even in the face of mounting evidence of health damage. It was also clear that errors had been made during the testing program.[20] This was the first real evidence of possible government negligence in the conduct of atomic testing at the NTS.

In 1980, a congressional report[21] summarized the findings of the hearings. The report denounced the AEC for its failure to warn downwind communities prior to atomic tests in spite of information known at the time about the health hazards of radioactive fallout, for its failure to accurately monitor radiation exposures downwind or to protect downwind residents, and for its knowing disregard of evidence of the questionable accuracy of the government's radiation measurements and radiation standards at the time. The report, referring to NTS downwinders as America's "forgotten guinea pigs," concluded that exposure to fallout from atomic testing was more likely than not the cause of cancer and other health issues faced by NTS downwinders.[22]

Further Epidemiological Proof of Fallout Harm

The disturbing findings of Lyon's 1979 leukemia study had created major controversy. In early 1981, a federal Radiation Research Committee concluded that the excess of childhood leukemia in southern Utah remained "unexplained on grounds other than possible fallout exposure."[23] Efforts by Lyon and his team to follow up on the 1979 study, to obtain better dose estimates for study participants, and to study a larger fallout exposed cohort were unsuccessful until Sen. Orrin Hatch (R-UT) intervened.

In 1982, with funding from three federal agencies, the NCI, DOE, and US Department of Defense (DOD), Lyon and his team were finally able to follow up their earlier study and the suppressed Weiss study to investigate whether the excess of leukemia deaths in southwestern Utah found in the 1965 and 1979 studies was actually related to exposure to fallout from atomic testing. Lyon and his team used estimates of radiation exposure in fifty-seven Utah communities published in the journal *Science* in 1984.[24]

Representatives from the DOE and DOD performed regular site visits and exerted substantial influence on the study's design, progress, and final interpretations.[25] The study nevertheless found a 7.82-fold increase in leukemia deaths among those less than nineteen years old in southwestern Utah during atomic testing.[26] Lyon found a statistically significant excess of leukemia deaths in northern Utah as well, using DOE estimates of radiation exposure. When these results were shared with the DOE, the agency revised NTS fallout exposure estimates for northern Utah communities downward, attributing greater percentages of the exposures to fallout from Russian nuclear tests. This caused the association of leukemia with NTS fallout in northern Utah to become insignificant. Lyon's team was denied access to the classified data used by the DOE to reassess doses.[27]

In 1984, Dr. Carl Johnson conducted a survey focused on cancer incidence in a cohort of Mormon families living in 1951 in towns with heavy fallout, including St. George, Parowan, Paragonah, and Kanab, Utah; Fredonia, Arizona; and Bunkerville, Nevada. Johnson compared his results with statistics he had gathered on nonexposed Utah Mormons across the state. Johnson found an excess of cancers and leukemias in the study group and attributed the excess to fallout exposures. He examined smoking, occupational history, and industrial carcinogens but found no other probable causes for this excess.[28]

In 1985 and 1986, during Phase 2 of the NTS fallout thyroid studies, Lyon and his team reexamined the cohort they had identified in 1965. A dosimetry model was created, and previous dose estimates for the cohort were corrected.[29] After the team corrected the doses and updated the disease classifications for a number of participants, the association between estimated radiation dose to the thyroid from atomic testing and subsequent thyroid neoplasms and thyroiditis among those exposed as young children was stronger.[30] The results of Phase 2, published in 1993,[31] identified a threefold excess of thyroid neoplasms among members of the study cohort with the heaviest exposures to NTS fallout—the first finding to suggest that people exposed as children to radioactive iodine are at higher risk of developing thyroid neoplasms. Despite this finding, the NCI issued an official press release deeming the results of the study "inconclusive."[32]

In October 1987, Lyon submitted a grant application to the National

Institutes of Health for follow-up examination of the former cohort of schoolchildren. The children, now grown, were just reaching the highest risk period for development of thyroid cancer, forty years of age and above. The grant application and a repeat application in 1991 were denied funding.

Fifteen years later, using newly corrected dose estimates and disease outcomes from the earlier phases of the thyroid study, Lyon and his team finally obtained funding to reassess the risk of thyroid neoplasms. Estimates of thyroid exposure dose from radioactive iodine in NTS fallout for representative persons had been made public by the NCI in 1997.[33] The NCI had developed a calculator to determine the dose received from NTS I-131 fallout for every person in all contiguous counties within the continental United States.[34] Lyon and his team used the NCI dose calculator to assign I-131 thyroid doses to the study participants. The calculated risk of thyroid cancer in the highest-dose group increased with the use of the newly corrected exposure dose information, leading to the conclusion that people exposed to radioactive iodine as children have an increased risk of thyroid neoplasms and autoimmune thyroiditis up to thirty years after exposure.[35] This was the first indication that very long latency periods can occur following exposure to radioactive iodine before thyroid cancers develop.

Bulloch II (1982)

Congressional hearings in 1979 on the health effects of fallout from atomic testing had uncovered indications of a "cover-up" by the government during Bulloch I in 1956 with regard to the effect of radiation on animals. Declassified AEC records obtained during the hearings revealed that "dose levels were almost 1,000 times the permissible levels for human beings" at the test site.[36] Prompted by these disclosures, six of the Utah sheep ranchers who had been plaintiffs in Bulloch I asked the district court to set aside its judgment in the case, alleging that fraud had been committed upon the court.

In May 1982, Judge Christensen, who had presided over Bulloch I twenty-six years earlier, heard four days of testimony. The plaintiffs proved that the AEC had known since 1951 that fetal lambs could be damaged by I-131 ingested by their mothers, based on the findings of the Hanford I-131 sheep studies led by Dr. Leo K. Bustad. The plaintiffs argued that the results of the Hanford sheep studies demonstrated that the AEC understood at the time of *Bulloch I* that the radiation health damage seen in Hanford's experimental ewes and lambs was identical to that seen in fallout-exposed Utah ewes and lambs that died in 1953.

In his testimony in *Bulloch II*, Bustad asserted that he had failed to make it clear to the Court in *Bulloch I* that some Hanford experimental sheep had shown symptoms identical to those in sick and dead Utah sheep, as it would have confused the court because all the Hanford lambs that had died had come from dams that had already become hypothyroid, and none of the tissue samples from Utah ewes that had been examined had been found to be hypothyroid.[37] Bustad's assertion was false as well as misleading. Hanford's I-131 sheep studies had shown that ewes fed 135 μCi/d of I-131 were not clinically hypothyroid at the time of lambing, whereas their lambs were severely hypothyroid at birth and did not survive for more than a few days.[38] Furthermore, the Utah tissue samples that Bustad had examined were not from the ewes that had died and were therefore not representative of the dead ewes.[39]

The plaintiffs proved that the AEC had pressured one of the veterinarians who had investigated the 1953 Utah sheep deaths to change his sheep autopsy reports to conclude that fallout had nothing to do with the sheep deaths. The plaintiffs also demonstrated that the AEC had provided misleading information on radiation dosages along with misleading answers to interrogatories. At the time of *Bulloch I*, the AEC had controlled all the data on where fallout clouds from atomic testing had traveled and on radiation measurements made by AEC off-site monitors. In *Bulloch I*, the AEC had revealed only some of the information it held and reported other information in a way that was deceptive, thereby misleading the court.[40]

Judge Christensen decreed that the government had used "improper means" that were "unacceptable as part of the judicial process" and that "clearly and convincingly demonstrate a species of fraud upon the court for which a remedy must be granted even at this late date."[41] He vacated his 1956 judgment, ordered the government to pay costs, and scheduled a new trial.

A three-judge panel of the Tenth Circuit, known for its extreme progovernment stance,[42] struck down the ruling, finding no evidence of fraud and accusing Judge Christensen of "abuse of discretion."[43] The Tenth Circuit later granted reargument en banc (before the entire Tenth Circuit). On May 22, 1985, the Tenth Circuit once again ruled against the sheep ranchers. This was the first time in the history of the Tenth Circuit that the court had overturned a district court's grant of a new trial. The ranchers filed a petition for certiorari to the US Supreme Court, which the court denied by only one vote.[44]

Hanford's Role in *Bulloch I*

Hanford scientists played a key role in *Bulloch I*, supporting the AEC's successful obfuscation of the link between fallout from atomic testing and the gruesome deaths of ewes and lambs in Cedar City. The outcome of *Bulloch I* hinged almost entirely upon the way the findings of Hanford's I-131 sheep study were represented in *A Comparative Study of Hanford and Utah Range Sheep* (HW 30119) (hereafter referred to as *A Comparative Study*), prepared for the AEC by Leo Bustad. The portrayal of the study findings in *A Comparative Study* is disturbing in light of what was actually known at the time from the Hanford sheep experiments. In the congressional hearings in 1979 that first revealed a possible cover-up of the link between fallout and Utah sheep deaths by the AEC, Dr. Harold A. Knapp, formerly with the AEC office of fallout studies, alleged, "The Hanford lambs exhibited symptoms similar to those of the Utah lambs" and added, "Yet, these results were not explicated in the text of the Bustad report [*A Comparative Study*]; rather, they were obscurely inferred through the listed references to the report."[45]

Bustad and his coauthors maintained that "radioactive iodine . . . must be administered in quantities in excess of 480 μCi/d to adult sheep in order to produce an impairment of health."[46] It is unclear whether this assertion refers to the health of adult sheep or their newborn lambs. If the assertion refers to impairment of health in lambs, it contradicts the results from the Hanford sheep experiments in 1951 showing thyroid damage in lambs of ewes fed I-131 in a dose as low as 45 μCi/d.[47]

Bustad claimed, "Premature lambing described in some Utah sheep flocks was not observed in the experimental sheep at Hanford although some test animals were fed up to 1800 μCi/d for 420 days."[48] This conclusion is false as well as irrelevant to the Utah sheep inquiry: it is false in that a number of Hanford experimental ewes fed 240 µCi/d gave birth prematurely,[49] and it is irrelevant in that prematurity was not an issue in the Utah sheep, whose lambs were born stunted or with grotesque deformities but not premature.[50]

Finally, Bustad asserted, "The large number of deaths observed in adult Utah sheep was not observed in any acute or subacute Hanford experimental sheep fed up to 280,000 µCi/d I-131 in a single dose. Deaths observed in experimental sheep did not occur until over five months following initiation of I-131 feeding and was attributed to systemic disease complicating the chronic hypothyroid state."[51] It is difficult to know what the reader is to conclude from this statement. Twelve percent of adult Utah sheep died, while all Hanford sheep that ingested 280,000 µCi of I-131 in a single dose died, but not before five months after feeding. One-tenth of this dose was shown in the Hanford experiments to kill all fetal lambs.[52]

Knapp believed that "the unusual and unexplained deaths of thousands of sheep in areas downwind of the Nevada Test Site in the spring of 1953 can be attributed to the fallout from eleven nuclear tests conducted between March 17 and June 4 of that year."[53] Knapp alleged that Bustad and his colleagues, collaborating with the AEC, had deliberately suppressed some of the findings of the Hanford I-131 sheep experiments in A Comparative Study while emphasizing others, creating a misleading impression.[54]

Knapp argued that because he himself had been misled by the Hanford sheep study results as presented in A Comparative Study, Judge Christensen and Dan Bushnell, attorney for the sheep ranchers, might also have been misled: "If someone were trying to hide something, one safe way to do it, and one which might help the conscience a little, would be to hide it in the references."[55] In Knapp's opinion, the deaths of ewes and lambs in the Hanford experiments were "magnificently disguised"[56] in A Comparative Study. Although the plaintiffs and the court in Bulloch I might have been able to access the information contained in the documents referenced at the conclusion of A Comparative Study, several of which were classified as restricted, this "in no way excuses you [Bulloch and colleagues] or the AEC from being forthright in the text and conclusions of your 1953 report."[57]

Bustad accused Knapp of trying to discredit his work. Knapp responded, "I am not trying to discredit your scientific work. . . . But, from what you tell me, I find it difficult to avoid a conclusion that the AEC deliberately and effectively withheld critical information on the effects of radioiodine on sheep, especially fetal lambs, and that it failed, for whatever reason, to give the sheep death problem a fair and competent evaluation."[58]

Downwinders Turn to the Courts

Communities downwind of the NTS now understood that cancers and deaths within families and in friends were likely the result of the negligence of the US government and its contractors in the conduct of atomic testing.

A number of NTS downwinders turned to state courts, seeking compensation for thyroid cancer, leukemia, and other illness through personal injury tort suits filed against private contractors that had conducted atomic testing for the government at the NTS. The contractors included Lawrence Livermore Laboratory and Los Alamos Laboratory, both operated by the University of California; Sandia Laboratory, a subsidiary of American Telephone and Telegraph; and Reynolds Electrical and Engineering Company.[59] State court offered the downwinders the advantages of jury trials and the ability to seek punitive damages, neither of which are available in federal court in suits filed against the United States under the FTCA.

Atomic testing contractors had expected these personal injury suits from the downwinders for some time. The US government and its contractors understood even before atomic testing began that the contractors might be sued for personal injury and property damage. The contractors did not have the protections of the sovereign immunity defense available to the federal government under the "discretionary function" exception to the FTCA.[60] However, the contractors had the protections of the "government contractor" defense, a federal common law defense immunizing an independent contractor employed by the federal government from liability when that contractor acts at the direction of the government. NTS contractors had the additional protection of indemnification agreements that they had signed with the US government, in which the United States agreed to pay attorneys' fees and any judgments relating to claims against the contractors for injuries from the US atomic testing program.[61]

The Warner Amendment

In spite of these comprehensive legal protections, the atomic testing contractors wanted more: they wanted to be shielded through retroactive immunity. They wanted to avoid litigation entirely. With the assistance of the Reagan administration, the contractors asked for help from Congress. The first attempt in 1983 failed when a congressional subcommittee found that the statute that the contractors had requested would be an unconstitutional interference with the vested causes of action already filed in state courts by a number of NTS downwinders.[62]

The following year, at the request of the Reagan administration, Sen. John Warner (R-VA) introduced the contractors' proposed statute as a rider (amendment) to the DOD's pending authorization bill, and it passed without a public hearing. Neither the House nor the Senate Judiciary Committees ever saw or considered this amendment.[63] The Warner Amendment was signed into law by President Reagan on October 19, 1984.[64] Challenges to the constitutionality of the Warner Amendment and later attempts to amend it have been unsuccessful.[65]

The Warner Amendment made the US government rather than atomic testing contractors the sole defendant in all pending and future lawsuits involving atomic weapons testing. The Warner Amendment was applied retroactively, resulting in the removal to federal court of the personal injury tort actions previously filed in state court against the contractors by NTS downwinders, deeming these to be tort actions against the United States.[66]

Had the NTS downwinders' personal injury suits been allowed to proceed in state court, during the discovery phase of litigation, the plaintiffs could have compelled the production of records revealing what the contractors and the government knew about the hazards of radiation and fallout at the time of testing in addition to information about the atomic testing program that is still classified today. If the downwinders' cases had gone to trial in state court, much of the information gained during discovery would have become part of the public record.

With the passage of the Warner Amendment, the only remaining avenue of legal recourse for the NTS downwinders was to file personal injury tort suits under the FTCA[67] against the United States in federal district court.

Irene Allen v. United States

Claims filed under the FTCA are heard by a federal court judge sitting without a jury.[68] Judge Bruce H. Jenkins of the US District Court, District of Utah, was randomly assigned to the case. The claims of 1,192 NTS downwinders were consolidated (combined into one proceeding) as *Irene Allen v. United States* (hereafter referred to as *Allen*).[69] The *Allen* plaintiffs asserted negligent release of radiation, failure to adequately monitor fallout released off-site, and failure to inform downwinder plaintiffs of danger.[70] Attorneys defending the government asserted that the AEC had acted within the scope of its legal discretion under the FTCA and that test site operators had no duty to warn the public about the dangers of atomic testing.[71]

Irene Allen was the first downwinder plaintiff, alphabetically and therefore lent her name to the litigation. Due to the high number of plaintiffs involved, twenty-four bellwethers (representative plaintiffs) were randomly chosen to represent the entire group. Each bellwether asserted that radiation-caused cancer or leukemia was "more probably than not"[72] the result of exposure to radioactive fallout downwind of atomic testing at the NTS. The outcomes of the bellwether cases would set a precedent for the resolution of the claims of the other plaintiffs.

The trial began in the fall of 1982 and lasted almost three months. On the witness stand, the bellwethers chronicled cancers and other diseases they believed had been caused by exposure to fallout from atomic testing at the NTS. Some of the plaintiffs broke down on the stand, tearfully recalling the cancer deaths of family members.[73] Others described radiation-related disability that had reduced or virtually eliminated their ability to earn an income, leading to lives of poverty. News reports of the bellwethers' leukemias; Hodgkin's disease; lymphoma; and cancers of the lung, stomach, brain, skin, uterus, ovaries, breast, pancreas, kidney, bladder, colon, prostate, thyroid, and other organs[74] created widespread public awareness of the human toll of atomic testing at the NTS.

Some of the bellwethers recalled family outings to watch predawn atomic tests, with beautiful red-and-orange mushroom clouds rising into the stratosphere. Often, pink or gray clouds passed overhead following the tests, from which a kind of (radioactive) snow would fall.

Toxic Torts and the Downwinders

The claims filed by the downwinders were referred to as "toxic tort" personal injury claims. A tort is an act or omission that causes injury or harm to another and amounts to a civil wrong for which the courts impose liability.[75] A toxic tort is a unique type of tort injury caused by exposure to a toxin such as a chemical or radiation.

In tort litigation, the plaintiff bears the burden of proving that the defendant's action caused the plaintiff's injury or illness. Toxic torts often entail exposures that are followed by decades-long latency periods before first symptoms of radiation-related disease appears. An ordinary tort action might involve A firing a gun at B, seriously wounding B. In this scenario, the cause of the injury to B is determinate (precisely defined) and direct. But with toxic torts, causation is far less direct, often referred to as "indeterminate," as in a case in which A irradiates B, who develops a tumor twenty-two years later.[76] During the extended latency period between exposure and disease, other conceivable causes for the plaintiff's disease may have occurred, creating an almost insurmountable hurdle for the toxic tort plaintiff who must prove that it was the defendant's actions rather than anything else that caused his or her injury.[77] Additionally, radiation-caused cancers cannot always be distinguished from cancers that occur without radiation exposure.[78]

Proving Factual Causation

In traditional tort litigation, the plaintiff must prove both factual (cause-in-fact) causation and proximate (legal) causation. The plaintiff must demonstrate factual causation on a "more probable than not" basis—that is, he or she must demonstrate a probability of causation greater than 50 percent.[79] Factual causation is the focal point of toxic tort litigation. Under traditional tort theory, proof of factual causation is met by either the *but for* or the *substantial factor* test. The but for test is met when an event or injury would not have occurred but for an act or omission by the defendant.[80] The more lenient substantial factor test is met when two or more causes join together to bring about an injury or event, and any one of the causes, operating alone, would have been the but for cause sufficient to bring about the injury or event. In this situation, each cause must be responsible for the event or injury. In most states, in toxic tort litigation, substantial factor causation is considered sufficient to establish factual causation.[81] Proximate (legal) causation is a separate policy-based concept applied to keep the scope of tort liability within reasonable bounds.

Judge Jenkins required NTS downwinder plaintiffs to prove factual causation under the substantial factor test, a more lenient test than the but for test later required of Hanford downwinder plaintiffs in *In re Hanford Nuclear Reservation Litigation*.[82] This meant that while the Hanford downwinders were required by the court to prove that but for the government's release of radioactive fallout from Hanford chemical separations, they would not have been injured, Judge Jenkins required the NTS downwinders to prove that the government's negligence in the conduct of atomic testing had been a substantial factor in causing their injuries.

Under the substantial factor standard applied by Jenkins, the court could reasonably conclude that an individual bellwether's cancer or other disease had been caused by exposure to radioactive fallout from atomic testing, absent persuasive proof to the contrary by the government.[83] Jenkins viewed statistical evidence as a guide rather than an answer.

Judge Jenkins believed that any kind of but for analysis to determine factual causation was problematic at best, particularly when a number of potential causes for radiation injury, including fallout exposure, background radiation, and medical X-rays, existed.[84] "But for" factual causation requirements have been called "an unrealistic burden to toxic exposure plaintiffs resulting in the dismissal of meritorious toxic tort claims."[85]

In toxic tort litigation, a relatively new area of tort law, courts have identified two categories of factual causation: general and specific causation. General (also called generic) causation refers to whether the toxin is *capable* of causing the injury claimed by the plaintiff. Specific (also called individual) causation refers to whether the toxin *more likely than not* caused the specific injury to the plaintiff asserting that injury. Courts often rely on expert opinion along with epidemiological evidence, employing statistical analysis and mathematical or computer modeling, to aid in determining factual causation. Toxic tort litigation frequently involves dueling experts and admissibility rules of evidence that determine which evidence is ultimately considered by the trier of fact.

Once the NTS downwinder plaintiffs had proven by a preponderance of the evidence that fallout from atomic testing was capable of causing their cancers and other illnesses, Judge Jenkins shifted the burden of proof to the government to prove that its negligence in the conduct of atomic testing had not caused the plaintiffs' injuries. Jenkins did not specifically use the terms "generic" and "specific" causation in his analysis. He concluded that the government's negligent failure to adequately monitor radioactive fallout exposures to downwind communities, combined with its failure to warn the public of fallout radiation risks and to advise the public on how to protect itself from harm, warranted shifting the burden of proof of causation from the downwinders to the government.

In his opinion, Jenkins cited cases in which a plaintiff had been injured but had no means of identifying the specific cause-in-fact of the injury. In these cases, the burden of proof was sometimes placed on the defendant to establish the factual details of the incident in question and to show that the defendant's conduct had not contributed to the victim's injury.[86] Shifting the burden of proof from the plaintiff to the defendant in this situation, Jenkins wrote, "reflects a sound application of important legal policies to the practical problems of trying a lawsuit: where a strong factual connection exists between defendant's conduct and the plaintiff's injury, but selection of 'actual' cause-in-fact from among several 'causes' is problematical, those difficulties of proof are shifted to the tortfeasor, the wrongdoer, in order to do substantial justice between the parties."[87]

In determining whether the government had met its burden of proof, the substantial, connecting factors considered by the court included the probability that the plaintiff had been exposed to ionizing radiation downwind of the NTS at levels above background radiation levels, whether exposure to ionizing radiation was recognized through epidemiological studies to cause the plaintiff's cancer, and whether the plaintiff had resided in geographic proximity to the NTS for some time between 1951 and 1962. Judge Jenkins also considered the plaintiff's age at the time of exposure; the known sensitivity of certain organs to radiation; each plaintiff's internal or external exposure dose as calculated by both plaintiffs' and defense experts; whether the latency period prior to the appearance of each bellwether's cancer corresponded to the recognized latency period for these cancers following exposure to low-dose ionizing radiation; and, based on epidemiological studies of NTS downwinders and similarly exposed populations, whether the statistical incidence of the plaintiff's cancer or other disease had been determined to be increased.[88] Judge Jenkins also referenced a report by the National Academies of Sciences concluding that thyroid cancer, breast cancer, and leukemia have very high sensitivity to radiation.[89]

The Trouble with Inexact Doses

The *Allen* bellwethers were each assigned estimated organ doses from internal and external exposure to fallout.[90] In his opinion, Judge Jenkins discussed the challenges confronting the DOE's Off-Site Radiation Exposure Review Project (ORERP) in its efforts to reconstruct doses to downwind populations: "The dependence of those activities upon assumptions, models and data from other times and places highlights both the inadequacy of the data regarding gamma exposure and the paucity of information concerning other exposure parameters."[91]

The AEC's failure to monitor individual exposure doses for people downwind of the test site became painfully evident during congressional hearings held on the atomic testing program. Humiliation over these deficiencies motivated the DOE to initiate ORERP in 1979, spending millions of taxpayer dollars in an attempt to reconstruct internal and external exposure doses to twenty-two target organs for those downwind of the NTS. ORERP was directed by the DOE, with staff drawn from Los Alamos, Livermore, and other national laboratories and researchers under contract with the DOE. Many of those involved had been employees of the AEC.[92]

Members of an advisory group appointed by the DOE to lend credibility to the ORERP project were troubled over the lack of accurate measurements of radiation exposure. They expressed concern about the large degree of assumption, uncertainty, and potential error inherent in the project due to the lack of basic empirical measurements.[93] An ORERP project scientist observed, "Had the offsite monitoring program taken adequate measurements of beta exposure, beta/gamma ratios and perhaps even regular measurements of alpha activity in dust particles that might be inhaled, the current estimation efforts would require fewer assumptions."[94]

The AEC's failure to monitor fallout exposure doses to civilians downwind would make it extremely difficult for the downwinders to prove in court that exposure to fallout had caused their specific cancers or other injuries. NTS off-site radiation monitoring personnel reported that fallout readings taken at the time of testing were not always accurate, varying substantially depending on whether the readings were taken at ground level, waist level, or in the bushes. As the winds blew through areas downwind of the NTS, fallout was concentrated or dispersed by mountains and valleys. Exposures downwind of the test site, much like exposures downwind of Hanford, were repeated and uneven, making the estimation of cumulative exposure dose much more complicated than for single-period exposures, as in Hiroshima, Nagasaki, the Windscale nuclear accident in Great Britain, or the Three Mile Island nuclear accident.[95] It was also difficult to accurately account for individuals' location during the entire period of atomic testing. All of these factors made accurate dose reconstruction a dubious undertaking.

Government attorneys urged Judge Jenkins to accept the ORERP exposure doses for the bellwethers as the probable upper limits of exposure and risk. Judge Jenkins refused to do so, instead considering the ORERP dose estimates to be minimum estimates.

For the bellwethers, the estimation of the fallout exposure dose to specific organs and organ systems varied substantially, depending upon whether dose calculations had been carried out by the downwinders' experts or the government's experts.

At trial, three of the bellwether plaintiffs were unable to establish that their cancers had been shown by epidemiological studies of other radiation-exposed populations to be caused by ionizing radiation.[96] For each of the remaining twenty-one bellwethers, Judge Jenkins evaluated whether the individual ORERP target organ dose or the alternate target organ dose calculated by plaintiffs' experts for that person was high enough, based on epidemiological studies,[97] to prove that that person's cancer was more likely than not caused by exposure to fallout. In his Memorandum Opinion in *Allen*, Judge Jenkins included a table[98] of the estimated doses to target organs for each bellwether calculated by experts for both sides in the litigation.

Judge Jenkins considered estimated fallout exposure dose to be just one of the substantial factors in the bellwethers' ability to prove causation. Jenkins's weighing of substantial factors to determine causation was far more equitable to the downwinders than the court's sole focus in *In re Hanford* on HEDR government dose estimates for bellwethers with thyroid disease and thyroid cancer.

Judge Jenkins weighed whether each bellwether's exposure to NTS fallout was a substantial factor in that person's cancer, examining whether the bellwether's exposure to fallout materially exceeded radiation exposure from other sources, including background radiation, medical X-rays, or other radiation treatments. When, by a preponderance of the evidence, it appeared that the government's conduct in carrying out atomic testing significantly increased the bellwether's risk of cancer, and that cancer was of a type shown by epidemiology and other sources to be caused by ionizing radiation, the inference could be rationally drawn by the court that the government's conduct was a substantial factor contributing to the bellwether's injuries.[99] Jenkins's decision expanded tort law, reducing the plaintiff's burden of proof where the defendant's actions were particularly objectionable and permitting plaintiffs to recover by showing possible causation rather than probable causation.[100]

In his Memorandum Opinion, Judge Jenkins wrote that the ORERP dose estimates did not seem to take into account the significant differences reported by off-site radiation monitors in measurements made in the open, as a fallout cloud passed over, and simultaneous readings of radiation deposited in the same location on the ground.[101] Expert testimony during trial revealed that off-site radiation measurements were made only at a limited number of locations, usually along highways. These scattered measurements did not account for possible hot spots of higher radiation. Judge Jenkins understood that the lack of accurate exposure dose information was a true disadvantage to the downwinders, commenting that in his opinion, "that so much [dose] reconstruction is now required says a good deal about the thoroughness of the original effort."[102] An accurate exposure dose was required to prove that a bellwether's exposure to fallout materially exceeded exposure from other sources. "The dearth of dose data made impossible the construction of standard radiogenic cancer cases by the downwinders against the government. These data do not exist because the government negligently failed to gather these data."[103]

Jenkins wrote, "Had the government accurately monitored the individual exposures in off-site communities at the time of the tests, accurate estimation of actual dosage for individual persons could have been achieved. The need for particular precautions could have been evaluated with confidence."[104]

Following the conclusion of the trial, Judge Jenkins and his clerk worked for more than seventeen months to craft the 489-page opinion, handed down on May 10, 1984. In his opinion, Judge Jenkins stressed that "the law imposes a duty on everyone to avoid acts in their nature dangerous to the lives of others."[105] Not only was Jenkins's landmark opinion based on a thorough review of the basics of nuclear physics, but it also broke new ground in the analysis of the scope of sovereign immunity provided by the discretionary function[106] exception to the FTCA, the subject of continued interpretation by the courts.

Jenkins determined that the delegation of authority to low-level AEC decision makers in the conduct of atomic testing did not fall within the discretionary function exception to the FTCA.[107] Jenkins ruled in favor of nine NTS downwinder bellwethers, holding the US government liable for illness and wrongful death from childhood leukemia, breast cancer, and thyroid cancer. He awarded $2.66 million, to be distributed based on losses suffered.[108] Jenkins found as a matter of law that the evidence was insufficient with regard to the other bellwethers to demonstrate that the defendants' negligence proximately caused the diseases claimed.[109]

Allen was the first ruling by a federal judge against the US government in a tort liability suit involving injuries to civilians from exposure to low-dose ionizing radiation. Jenkins concluded that the government had been negligent and that certain plaintiffs' cancers were more likely than not caused by exposure to nuclear fallout. This was the first time atomic testing had been determined by a federal court to have caused cancers, and the decision was hailed as a "landmark ruling" that could open the door to recovery for others exposed downwind of the NTS.

Reversal of *Allen*: The Implications for Hanford Downwinders

The government immediately appealed Jenkins's ruling. As it had in *Bulloch II*, the Tenth Circuit exerted its power against the interests of the downwinders, reversing plaintiff verdicts without hearing the evidence. The circuit court ruled that the government could not be held liable under the FTCA because the atomic testing program's public information decisions were a discretionary function, one of the exceptions to the federal government's waiver of sovereign immunity.

In his concurring opinion, Circuit Judge Monroe McKay expressed sympathy for downwind cancer victims who "have borne alone the costs of the AEC's choices."[110] Judge McKay argued that the courts had no choice but to deny compensation to these victims until Congress either amended the discretionary function exception to the FTCA or passed a specific compensation bill.[111]

The downwinders appealed to the US Supreme Court, which in January 1988 refused to hear the case.[112] Denial of certiorari[113] by the US Supreme Court in the NTS downwinders' case was detrimental to the Hanford downwinders as well, as it led to a decision by a prominent Seattle law firm not to represent the Hanford downwinders in upcoming litigation against the government for injuries from radioactive fallout released from Hanford. "We've stopped our investigation. The *Allen* case was an imposing decision," said Andrea Brenneke, a legal assistant at Schroeter Goldmark & Bender,[114] the Seattle firm well known for representing toxic tort exposure plaintiffs in complex litigation.

"It's a tragedy the government can't be held liable," Brenneke said. "Our partners think people were harmed by Hanford. But as long as the law is that the government can do anything it pleases and not be held accountable, victims will have no recourse."[115]

Hanford downwinder Tom Bailie said he was "deeply disturbed about the decision to suspend the investigation. I'm slowly finding out that Washington State's downwinders are finding themselves in the same category as the Nevada downwinders—where there's no legal recourse and the US Department of Energy is above the law."[116]

The Supreme Court's denial of cert in the *Allen* decision came just as other major plaintiffs' law firms, several of which had represented downwinders following the Three Mile Island nuclear accident in 1979, were meeting in late January 1988 to decide whether to represent Hanford downwinders in upcoming litigation. The firms were under pressure because the two-year "discovery rule" applied in toxic tort cases, which required that the Hanford downwinders' claims be filed by the end of the following month—two years from February 27, 1986, the date of the DOE's mass declassification of historical Hanford documents that revealed Hanford's legacy of radiation releases downwind to the public.

8

By the mid-1970s, Juanita Andrewjeski and her husband, Leon, had begun to notice that something was very wrong within their rural farming community, located directly across the Columbia River from Hanford. Young men working the fields in the small towns of Ringold, Mesa, and Eltopia were suffering heart attacks at an early age. A number of the Andrewjeskis' neighbors had been recently diagnosed with cancer. Then, Leon was found to have heart disease.

Juanita wanted to figure out what was happening to her community and her family. Following Leon's diagnosis, she began to track heart attacks and cancers. She made a list of the names, ages, and addresses of afflicted neighbors, plotting the information on a map of the area. Every X on the map represented a heart attack, every O a cancer diagnosis. The number of Xs and Os grew to alarming proportions. The map and an emergency radio had recently been distributed by the Benton and Franklin Counties Department of Emergency Management to all families within a ten-mile radius of Hanford. Families were told that officials would automatically activate the radio in case of airborne radiation release from an accident in Hanford's only remaining operational reactor, the aging N reactor.[1] In the three decades that the Andrewjeskis had lived on the farmlands across from Hanford, this was the first time a government agency had warned them about the possibility of airborne radiation from the nuclear facility.

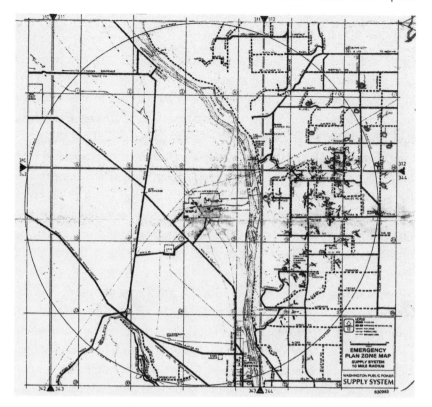

Juanita Andrewjeski's map of cancer and heart attacks in farmlands downwind of Hanford. Photo by author.

Death Mile

Tom Bailie, a friend of one of the Andrewjeskis' sons, contacted Karen Dorn Steele, a young reporter with the *Spokesman-Review* in Spokane, to tell her about the Andrewjeski map. Bailie later took Steele on a tour of an area close to his farm where nearly every family along a mile-long stretch of road had one or more members with cancer. That area became known as the "death mile." Steele began to report on the growing numbers of people in the area diagnosed with cancer and other disabling disease. Public concern grew about whether Hanford was somehow responsible for the increasing reports of illness. Andrewjeski's map was picked up by newspapers across the region as anecdotal evidence that something was very wrong downwind of the nuclear facility.

Hanford's Silent Harm

In 1983, the restart of the Hanford PUREX[2] plant and the government's plan to build a permanent national nuclear waste dump on the vast nuclear site triggered widespread public apprehension over Hanford as a nuclear neighbor. In Spokane, less than two hundred miles northeast of Hanford, Rev. Dr. William H. Houff presided over the congregation of the Unitarian Universalist Church. Houff was also a trained chemist and had long been troubled by the reckless use and misuse of radioactive elements. He became further concerned in February 1984 when he saw an article in the *Seattle Times*,[3] initiated by an anonymous whistleblower's call to the paper, reporting that radiation monitors had detected a large undisclosed release of radioactive thorium over several weeks' time from the restarted PUREX plant.

At a meeting of area ministers, Houff heard a speaker warn of the potential hazards of the planned permanent nuclear waste dump at Hanford. If the plans were approved, hundreds of shipments of nuclear waste would pass through Spokane on the way to Hanford. Houff felt a moral obligation to share his concerns with his congregation. He drafted a sermon entitled "The Silent Holocaust," which he delivered on May 20, 1984:

> Today, the word "holocaust" is most frequently associated with nuclear war . . . as in "nuclear holocaust." But the holocaust I'm looking at is not presaged by the mushroom cloud. Rather, it is silent. Not only is it silent, but except through certain kinds of scientific instruments and the broken health of living creatures, atomic radiation is not detectable at all. Its silence is also manifest in the fact that, despite massive evidence to the contrary, American officialdom, both in the private sector and the public, has uttered no words of alarm . . . but have, indeed, frequently taken extravagant measures to silence those who do sound a warning.[4]

Houff's sermon became a driving force behind the struggle to understand the dangers to the public posed by Hanford operations. About thirty people stayed after Houff's homily to ask questions. They returned several weeks later for presentations on how radiation can damage human health. In September 1984, Houff's congregants formed HEAL, a citizens' group that would study many aspects of Hanford operations. Early HEAL members included Dr. Al Benson, a Spokane chemistry professor who believed citizens could be empowered to make a difference if they understood the issues, and freelance journalist Larry Shook, who had written a number of articles on Hanford. Shook shared his research techniques with the group, teaching HEAL members how to track down relevant materials.

HEAL sponsored two educational forums in October 1984, the first examining the hazards of plutonium and the second featuring keynote speaker Dr. Alice Stewart, a British physician who spoke about the dangers of radiation exposure to children. The following month, HEAL invited the DOE's Hanford site manager, Michael Lawrence, to speak about potential hazards from Hanford to people living in the area. At an emotion-laden meeting, Lawrence tried to convince HEAL members that Hanford was safe while the audience disputed his assurances.

In 1984, Steele began to look into a possible connection between the deaths of sheep and newborn lambs downwind of the NTS in Cedar City, Utah, in 1953 and the 1961 deaths of sheep and newborn lambs within the farming communities downwind of Hanford. After months of research, she finally amassed enough information to persuade her editors at the Spokesman-Review that she was on to something and that Tom Bailie, with whom she had spoken several times, had a story to tell.

She went with Bailie to meet Nels Allison, who had lost many ewes and newborn lambs on his farm that night in 1961. They talked about the deformed lambs, the "little demons," and how Allison had nearly gone bankrupt from the loss of his animals. Steele asked Bailie and Allison if there was anyone else who had lost sheep and lambs whom she might talk to. Allison warned them that they were touching on a sensitive topic and that many area farmers would rather ignore Hanford than ask hard questions.[5] Steele then asked Bailie, "Tom Bailie, do you know anything about radiation and the effects on sheep at all? Do you really know what happened in Utah?"[6] Bailie said he knew little about the Utah case or about radiation generally. Steele told him that the deformed lambs born on Hanford-area farms were similar to the deformed lambs born in Cedar City, Utah, downwind of atomic tests at the NTS. Steele was now sure she was on to something big.

In 1985, HEAL, the Environmental Policy Institute of Washington, DC, and the *Spokesman-Review* filed broad Freedom of Information Act requests seeking Hanford operational records from the 1940s through the 1960s that were still classified on grounds of national security. The pressure on the DOE to declassify early Hanford operating records mounted.

"No Reason to Expect Observable Health Impacts"

On February 27, 1986, in front of an overflow crowd in Richland, Lawrence declassified nineteen thousand pages of early Hanford environmental monitoring reports and related records chronicling off-site releases of radiation and chemicals from the facility over more than forty years. At the accompanying press conference, Lawrence insisted, "There is no reason to expect observable health impacts" from the releases.[7] When questioned, other Hanford officials at the press conference conceded that no effort had been made to look for harm.[8]

CDC officials reviewed the declassified records and reported that Hanford's radioactive emissions were the largest ever documented from a US nuclear facility.[9] Millions of curies of radiation had been secretly released over forty years to the air and to the waters of the Columbia River.[10] Hanford downwinders, suffering from radiogenic diseases and cancers identical to those reported by NTS downwinders, were appalled that they had been unknowingly exposed to Hanford's radiation releases.

Steele, Spokane journalist Tim Connor, and others methodically combed through the mountain of declassified documents, many of which were faded with age to the point of indecipherability. Steele reported on what they found, which included documentation of off-site radiation emissions, both negligent and intentional, from the Hanford facility. One of the intentional releases revealed within the records was the Green Run experiment conducted in December 1949. Also among the information unearthed was the startling revelation that over 750,000 curies of airborne I-131 had been released throughout eastern Washington, Idaho, western Montana, northern Oregon, and southern British Columbia over four decades. In 1951, the failure of silver reactors (chemical reactants to remove iodine) in a reprocessing plant had resulted in the release of nineteen thousand curies of I-131.[11] From 1952 to 1954, much of eastern Washington had been showered with radioactive ruthenium particles, and during three months in 1957, an average of thirty-four thousand curies of radiation per day had been dumped into the Columbia River along with reactor cooling water.[12] An additional fifty thousand pages of records relating to early Hanford operations were declassified over the next five years. Among other revelations, these records showed that I-129, which has a half-life of sixteen million years, had reached some of the deep aquifers under Hanford and was present in wells in farms across the Columbia River.[13]

The governors of Washington and Oregon demanded to know whether and how these chronic radiation releases had affected the health of communities downwind and downriver of the Hanford facility. In mid-1986, the CDC convened the Hanford Effects Panel, an independent panel of scientists, Washington radiation officials, and representatives of area Indian tribes, to help answer this question. The panel found documentation within the declassified records that I-131 radiation doses to the thyroid for people living in the Hanford region in the 1940s were significant enough to have resulted in health problems and to have created special risk to infants and children, the most vulnerable members of the population.[14] The panel recommended dose reconstruction for the emissions during early Hanford operations, when radiation releases were at the highest levels, along with a thyroid cancer study of twenty thousand children who were born in Benton or Franklin County and who were younger than nineteen years of age within the period of 1944 to 1953.[15] The first recommendation resulted in the HEDR and the second in the HTDS.

In April 1986, just two months after the public first learned about Hanford's chronic irradiation of the people of much of the Pacific Northwest, the nation's attention was abruptly shifted to the catastrophic explosion of Chernobyl no. 4, an old-style graphite-moderated reactor similar to the Hanford reactors.[16] Both Chernobyl no. 4 and the Hanford N reactor were built of graphite blocks and were dual-purpose plutonium/power production reactors without containment domes.[17] The N reactor, commissioned in 1963, had been designed to last twenty years and was now past the end of its useful life.[18]

Following the Chernobyl nuclear disaster, Congress focused on hypothetical injuries to US populations should an accident occur in Hanford's Chernobyl-like N reactor.[19] It is conceivable that congressional hearings, akin to the series of congressional hearings that examined the health

effects of fallout from atomic testing at the NTS, would have been convened on the health effects of Hanford's fallout had the nation's focus at
the time not remained on the Chernobyl disaster and its implications for
the United States.

Dose Reconstruction and the Thyroid Study

The HEDR was the first government-sponsored project to reconstruct radiation doses to the public from a US nuclear weapons production facility.
The Department of Justice, tasked with defending Hanford contractors
under indemnification agreements signed in the early years of Hanford
operations,[20] had initially opposed dose reconstruction as useless "public
relations" but quickly changed its mind once the first suit for personal
injury from Hanford radiation was filed by members of the Colville Confederated Tribes in 1987.[21]

The governors of Washington and Oregon demanded a dose study conducted without influence from the DOE, but the DOE refused to pay for an
independent dose reconstruction project and instead began the HEDR in
1987 under the direction of Battelle Pacific Northwest Laboratory, a long-
term DOE contractor, with a group of independent experts referred to as
the Technical Steering Panel overseeing the work.[22]

The HEDR was touted by the DOE as unbiased and scientifically neutral. However, public concern over DOE's control of the HEDR resulted in
the transfer of project management to the CDC in 1992, three years before
completion of the project in 1995.[23]

Congressionally mandated in 1988, the HTDS was the only epidemiological study done to assess radiation damage in Hanford downwinders.[24]
The study began in 1991, and its preliminary findings were announced in
January 1999. The Fred Hutchinson Cancer Research Center in Seattle led
the study, with the CDC overseeing its work. The stated purpose of the
study was "to answer the question: 'Did exposure to iodine-131 from Hanford result in increased incidence of thyroid disease?'"[25]

The HTDS focused solely on thyroid damage following childhood exposure to I-131, one of a range of "biologically significant"[26] radionuclides
secretly released from Hanford as by-products of the chemical separation
of plutonium.[27] The government did not study the prevalence of cancers

far less survivable than thyroid cancer or the incidence of autoimmune disease, miscarriages, stillbirths, infertility, heart disease, and other potentially radiogenic diseases reported by Hanford downwinders. An official from the Washington State Department of Health revealed that "no other group of civilians in the world is known to have been exposed to as much radiation over as much time as residents of this [Hanford] region."[28] In spite of this stunning revelation, the downwinders were never given a say on whether additional epidemiological studies should be conducted to assess the prevalence of cancers and other potentially radiogenic diseases unrelated to the thyroid reported among populations downwind of Hanford.

The group of downwinders studied by the HTDS was restricted to those born between 1940 and 1946 to mothers who lived in seven counties in eastern Washington—Benton, Franklin, Adams, Walla Walla, Okanagan, Ferry, and Stevens—three of which, Ferry, Okanagan, and Stevens, were considered to be low-exposure counties.[29] The span of birth years in the HTDS cohort was far narrower than the span of birth years recommended for study by the independent panel convened by the CDC, which had advised that the study include a cohort of twenty thousand children younger than nineteen years of age in 1944 through 1953. I was born in Benton, one of the study counties, but was excluded from the HTDS, as I was born in 1950, outside the study cohort of those born between 1940 and 1946. My thyroid and parathyroid disease, my mother's hypothyroidism, my father's hypothyroidism, my mother's hyperparathyroidism, and my father's thyroid cancer were not considered in the findings of the HTDS, as we were all outside the defined HTDS study population.

Researchers tracked down birth certificates for 5,199 people and found that 4,350 of these people were still alive, and 849 had died, a higher-than-expected number of deaths in the population. Ultimately, 3,400 people participated in the study. The HTDS did not include an unexposed control group for a comparison of disease rates between the exposed and non-exposed populations. Instead, the study compared incidence of thyroid disease in those with higher HEDR doses and those with lower HEDR doses, looking for a "dose response," that is, higher incidence of thyroid disease in those with higher HEDR doses than in those with lower HEDR doses.[30]

Study participants were asked to complete dietary and residence history forms that reported where they had lived and spent their summer vacations as children during the period 1944–1957, more than three decades earlier, and to accurately report the number and size of servings of specific foods and dairy products they consumed on a daily basis. All participants were examined for the presence of thyroid disease. Following requests from the Hanford Downwinders' Coalition, thyroid ultrasounds were used to increase HTDS clinicians' ability to detect thyroid nodules. HEDR scientists, using computer programs and the information collected from the participants' dietary and residence history forms, estimated participants' I-131 exposures.

By the time preliminary results of the study were released in 1999, the downwinders had waited more than eight years since the start of the study, certain that the HTDS would validate their lived experience—thyroid disease and thyroid cancer that afflicted families, neighbors, and friends.

A communication embargo on the findings of the study was instituted by the CDC prior to the release of the preliminary results in January 1999, creating a cloak of secrecy that frustrated those who of us had followed the study since its inception. In spite of the embargo, the study's preliminary findings were revealed to the public in the morning edition of the New York Times on the day the results were to be announced. A staffer attending a congressional briefing on the study in Washington, DC, the day before had leaked the results to a New York Times reporter, resulting in banner headlines: "No Radiation Effect Found at Northwest Nuclear Site."[31]

The meeting to publicly announce HTDS preliminary findings was scheduled for the evening of January 28, 1999. Because people had learned of the negative conclusions of the study in that morning's New York Times article, they arrived at the meeting ready to argue with the Fred Hutchinson team. The downwinders already knew what the Fred Hutchinson team was going to say, and many were furious.

I attended the meeting. I couldn't stay away. As we entered the meeting room in a Richland hotel, we were handed a glossy pamphlet that just served to add fuel to the fire. The pamphlet contained a statement that was highly offensive to those who had lost family members and their own health to thyroid disease and cancer: "[HTDS provides] rather strong evidence that exposure at these levels does not increase the risk of thyroid disease. These results should consequently provide a substantial degree of reassurance to the population exposed to Hanford radiation that the exposures are not likely to have affected their thyroid or parathyroid health."[32] The statement, intended to reassure us, instead left us incredulous and many scientists and other observers skeptical. "The way in which the report was released showed a contemptible lack of sensitivity to the individuals whose personal well-being and family and community health have been, and continued to be, jeopardized by past exposures to Hanford's radiation."[33] An editorial appearing in the *Seattle Times* suggested, "The US Centers for Disease Control and Prevention deserves an 'F' for its presentation of the results of the Hanford thyroid study, a chorus of critics in the Northwest say."[34]

Sally Sanders of Kennewick had read the article in the *New York Times* that morning and had come to the meeting prepared to dispute the study's negative findings in her own way. She had lost two family members to thyroid cancer, and her brother had also been diagnosed with the disease. She sat through the entire meeting holding a sign above her head that read, in large block letters, "I DON'T BELIEVE IT." Angry audience members interrupted prepared remarks by Fred Hutchinson researchers. A long line of people waited at a microphone set up in the middle of the room, registering their objections to the study's findings. Widows and widowers from small towns in eastern Washington shared stories of spouses who had grown up downwind of Hanford and who had later passed away from thyroid cancer. People suffering from thyroid cancer, hypothyroidism, and hyperthyroidism angrily questioned how the CDC and Fred Hutchinson could reassure them that their thyroid health hadn't been harmed by Hanford's radiation. Several people asked why the study hadn't looked at the other cancers and autoimmune diseases reported by the downwinders. Some people complained that the study should have included people born in the 1950s, many of whom now suffered from thyroid disease and thyroid cancer. As the evening progressed, audience members grew angrier, while the study authors grew increasingly defensive.

The HTDS was designed as a retrospective cohort study that looked for a "dose response"—that is, whether thyroid disease occurred more frequently in study participants with higher HEDR doses than in participants with lower HEDR doses.[35] The CDC's preliminary report laid out

Sally Sanders of Kennewick protesting the "no thyroid harm" findings of the HTDS. Photo reprinted with permission of *Tri-City Herald*.

the study's rationale: "If there were a link between radiation dose and thyroid disease, we would expect thyroid diseases to occur more frequently in study participants with higher radiation doses."[36] A lot depended upon the accuracy of the HEDR doses, the source of a great deal of controversy.

Buried in the findings of the study was a "surprising amount of thyroid disease," including hypothyroidism in 27 percent of female participants,[37] although its prevalence was not [HEDR] dose related. The overall incidence of autoimmune thyroiditis of almost 19 percent, with the number reaching 24 percent for women in the study, was greater than might be expected from the results of normal population studies. These numbers were also higher than one might expect from other epidemiological studies of presumably normal populations.[38] The study additionally found neonatal deaths to be 20 percent higher than expected among downwind communities.[39]

The study's findings contradicted the lived reality of many Hanford downwinders. Somehow, according to the Fred Hutchinson researchers, more than 750,000 curies of airborne I-131 had been released from Hanford over forty years without harming the thyroids of fetuses, infants, children, and adults downwind. Why, then, had increased thyroid disease and thyroid cancer been found in studies of people exposed to fallout from US nuclear tests in the Marshall Islands, children in Utah and Nevada exposed to I-131 from NTS atomic tests, communities downwind of the Three Mile Island nuclear reactor accident (where only fifteen to twenty-four curies of I-131 had been released),[40] and people exposed to I-131 in fallout from the Chernobyl nuclear reactor explosion?

A columnist for the Seattle Times had the following reaction to the study findings:

We're 10 years older and $18 million poorer, but we still don't know whether Hanford downwinders were harmed by its radioactive emissions. We do know that twenty percent more of them are dead than expected. And we do know that Eastern Washingtonians were found to have two to three times more thyroid disease than other populations generally.

But those seemingly alarming findings may not mean a thing, according to researchers at the Centers for Disease Control in Atlanta and Seattle's Fred Hutchinson Cancer Research Center that studied the downwinders.

Then again, maybe the findings do mean something. No one, it seems, can say for certain.

The Hanford downwinder thyroid disease study is one of the most maddening chapters in the annals of epidemiology.[41]

The downwinders soon took matters into their own hands, insisting on an extended peer review of the study by a subcommittee of the Board on Radiation Effects Research of the National Academy of Sciences (NAS).[42] Two representatives of the downwinders traveled to Washington, DC, to meet with a highly placed CDC official, hand carrying a letter signed by representatives from more than twenty-two downwinder, Native American, and other citizens' groups in the Pacific Northwest. The letter expressed anguish over the study's purported conclusion that Hanford's massive releases of airborne I-131 had not caused harm to downwind populations and requested that the CDC provide funding for a review of the HTDS to determine whether it had been conducted properly and whether its findings were scientifically credible. In response, the CDC approved one of the first extended reviews of an epidemiological study to be requested by the population under study.[43] The review would concentrate not only on the scientific aspects of the study typically involved in a review by the NAS but also on the way the study's preliminary findings had been communicated to the public.

Why Did the HTDS Focus on Dose Effect?

The NAS peer review took nearly a year and included several public meetings within the Pacific Northwest. In June 1999 at one of the public meetings, held in Spokane, an Oregon state health educator with the HHIN made the following statement:

> I am a nurse and health educator for the Hanford Health Information Network. I have held this position for six and a half years and would like to share with you my experience with the downwinder population exposed from the Hanford Nuclear Site.
>
> In April of 1994, the Oregon Hanford Health Information Network

created and conducted an advertising campaign to encourage those persons exposed to radiation to call the Network for information regarding health effects. Information obtained from AT&T showed that over 11,000 phone calls were made to the 1-800 phone line in response to the campaign.[44]

The callers reported health effects, which they believed to be related to the radiation releases from Hanford. It is important to note that these thousands of callers were interested in information about their health effects and had no information at that time about the estimated dose of radiation they had received. They were calling because they had health effects that they believed were related to the Hanford emissions.

As we responded to their questions and concerns, a picture of the illnesses of this population began to emerge. Primarily we heard of thyroid disease: thyroid disease in twenty-year-olds, thyroid disease in a larger number of males than seemed usual, thyroid cancer in children, hypothyroidism with secondary effects to the reproductive system, the immune system, and the nervous system.

The callers shared that their thyroids were dysfunctional, that their thyroids had nodules, that their thyroids had been removed, and that they would have to take medication for the rest of their lives because their thyroids were not operable.

The dose of radiation to their thyroid was not part of our discussion because we had no information on dose at the time. The Hanford Environmental Dose Reconstruction Project did not release its findings until April 21, 1994, and only then for representative doses, not individual doses.

It is my opinion that chronic long-term exposure to iodine-131 in the air, in the water, in the soil, in the food, in the milk, in whatever dose, resulted in thyroid disease.

The Hanford Thyroid Disease Study was eagerly awaited by the downwinder population and me. When the results of the study were announced I was astounded! This study does not affirm my experience with the thousands of downwinders with whom I have spoken who call the Network to talk about their thyroid and other diseases.

I question why the study looked for a dose-related effect? The information I have from this population is that there is disease, with a wide variation in exposure and dose. Science may appreciate knowing how dose-response to disease was found by the Hanford Thyroid Disease Study. The downwinders I have spoken with know that the study does not reflect the disease they have experienced.

No study has been done with a population exposed to constant radiation in varying amounts over a long period of time. Neither has there been a study that can account for each individual response to a stimulus. Two people can stand in the sun. One person gets a tan. The other person gets burned. The effects of radiation to the body over time and in varying amounts has not been tested, and probably will never be, as who would consent to such an experiment?

It is not possible to say that the thyroid disease in this population is not related to the Hanford emissions. There are health effects in this population that the design of the study does not address. The downwinders are not reassured that the emissions from Hanford did not contribute to their thyroid disease.

With all due respect to the researchers, the results of the Hanford Thyroid Disease Study are not conclusive, and do not accurately reflect the numbers of persons with thyroid disease and other diseases in the Hanford population.

Thank you for your time.[45]

Plaintiff 17: Geneva Shroll

Geneva Shroll, born in Ritzville, Washington, in 1943, was a participant in the HTDS. Researchers from the Fred Hutchinson Cancer Research Center found nodules on her thyroid; biopsies indicated that they were not cancerous. Shroll was reported as a HTDS participant without thyroid cancer, one of two HTDS participants featured in this book who were found to be cancer-free at the time of the HTDS examinations. Several years after the negative findings of the HTDS were publicly announced, both were diagnosed with thyroid cancer.

Here is her story.

I was born in Ritzville, Washington, in 1943. I lived there until I married. Even after I married, I returned from time to time.

While growing up, we had our milk delivered from a local dairy. We had two plots of land where we grew a lot of berries and vegetables. I don't

Geneva Shroll as a child. Photo courtesy of Shroll family.

remember when I first heard about Hanford's radiation releases. I'm guessing it was not until the mid- or late 1980s.

I was part of the HTDS, as Ritzville was in the pathway of radiation releases. Researchers from Fred Hutchinson Cancer Center in Seattle came out to Spokane and set up in one of the hospitals. I believe there were three medical personnel who carried out the examinations. They performed an ultrasound and found six nodules on my thyroid. They asked permission to do biopsies of the nodules, and I agreed.

No cancer was found at that time. They said I wouldn't necessarily get cancer, but I did—years later, after the study was concluded. My husband was examined for the study too because he is also from Ritzville. He was found to be negative for thyroid issues.

In 2011, I went to a doctor for inflamed salivary glands. The doctor noticed a lump in my throat that I had been meaning to have checked but hadn't as yet had that done. I agreed to a biopsy that resulted in the finding of cancer in my thyroid. A thyroidectomy was scheduled immediately. Later, a large goiter was found in my chest. I had surgery again in 2012 to remove the goiter. I will receive follow-up care the rest of my life.

I had already signed on as a thyroid disease plaintiff in the Hanford downwinder litigation, but I immediately changed my status to thyroid cancer plaintiff.

My mother had cancer as well as heart issues, and my father had heart issues as well. My husband had surgery for kidney cancer in 2010. We will never know whether our exposures to Hanford's radiation caused these health issues.

I hope that the whole world has realized the horrors and extreme dangers of radiation and will not forget the lessons that history has shown us.

Plaintiff 18: Keith Lindaas

As was true for many people who grew up downwind of Hanford, the HTDS finding that Hanford I-131 did not cause harm to downwind populations did not reflect Keith Lindaas's reality. Keith grew up in the little town of Dayton, Washington, southeast of Hanford. In his twenties, he noticed a swelling in his neck. A biopsy later revealed thyroid cancer.

This is his story.

I was born in Colorado in 1938 but moved to Dayton, Washington, in the summer of 1945. Dayton is about forty miles southeast from Hanford as the crow flies. I lived in Dayton from 1945 until the summer of 1957.

Dayton was a small town of about twenty-five hundred. There was a creamery a couple of blocks from my house. We got all our milk from the creamery. As a child, I was smaller than other kids my age, so my parents thought I should drink milk. I liked it, so I drank quite a lot. We also ate a lot of fresh fruit and vegetables, as we had an extensive garden in our backyard.

Things went along okay until, in my twenties, shortly after I was married, I noticed a swelling in my neck. It looked something like the mononucleosis I developed in high school. The swelling would come and go. I went to the doctor, and he said not to worry about it. Then, I went to a naturopathic doctor who told me that if it didn't go down, it should be surgically removed. All this time, my wife kept after me to do something.

I was told I should have my wisdom teeth extracted to see if that took care of it, so I had that done. The oral surgeon said I should really have a specialist look at the swelling. I went to Swedish Hospital in Seattle and had a consult with an endocrinologist, who said I needed a biopsy. He said the swelling might be thyroid cancer. He told me that thyroid cancer normally appears in females and those of Mediterranean heritage, and since I was both male and Scandinavian, there was a 99.9 percent chance it was benign.

He did a biopsy and discovered it was thyroid cancer. By this time, we had four kids, the youngest only two years old. I was trying to support my family and was working full-time. The news was devastating, traumatizing. I don't remember whether we told the kids, but we did tell our parents.

The surgeon recommended that the entire thyroid be taken out. The surgery back then was different than now. The surgeon opened an entire flap of skin covering the front of the neck. Some of my friends say that this is my "Hanford necklace." Nowadays, the incision is much smaller. Also, now they give you radioiodine after the surgery and a CT scan to check for hot spots of cancer. They didn't do that back when I had my surgery.

I lost some of my parathyroid glands in the surgery. Another possible risk from the procedure was loss of muscle function if nerves were cut. I was very concerned about this, and my doctor extended the surgery to try to avoid this. As the result of this extra care, I still have full use of both arms, with no nerve damage.

Following the surgery, I began to take synthetic thyroid hormone, Synthroid or Levothyroxine, I don't recall which. I have adjusted ok. I consider myself fortunate that I didn't have any other side effects.

Later, when I needed follow-up testing, I thought about whether I should have the radioiodine and CT scan since I hadn't done so earlier. I decided to go through the procedure, and had to go off the thyroid meds for a period of time beforehand. My wife thought I was kind of grumpy while off the synthetic thyroid. The test, in 2001 or so, found some hotspots. Two or three doctors we consulted said to just watch these spots.

The only other health issues I have had is a skin melanoma that was just operated upon. I don't know if there is any connection to my exposure to Hanford radiation. It could be the result of sun exposure I got as a kid. This melanoma was just about a month ago. The surgeon ordered a CT scan since I had had thyroid cancer. The CT scan showed the nodules or hot spots still lit up. My wife and I continue to be worried about these hot spots.

It's kind of ironic that I would end up with a cancer associated with Hanford radiation releases. In the 1970s, as members of an antinuclear group, my wife and I volunteered to obtain signatures on petitions against Hanford. This was long before anyone knew of the radiation releases from the facility. It wasn't so much a protest against Hanford as it was an expression of concern over the dangers of nuclear waste. The long half-lives of much of the waste make it dangerous for thousands of years. I remember that we got a lot of flak from people on the street when we asked them to sign the petitions.

I think that the government should have owned up to what happened to those of us exposed to Hanford radiation downwind. It seems like they were just waiting for all of us to die off.

Something's Amiss with the HTDS

The HTDS finding that no thyroid health harm had been done to downwind communities from exposure to the I-131 in Hanford's fallout

Keith Lindaas. Photo courtesy of Lindaas family.

conflicted with other epidemiological studies that had found increased incidence of thyroid disease and thyroid cancer in children exposed to I-131 from nuclear fallout.[46] The National Research Council advised that "the incompatibility of HTDS findings with other studies of radiation and thyroid disease should be reexamined, taking into account the impact of dose uncertainties."[47] While the study was well designed, shortcomings in the analytical and statistical methods used by the study's investigators overestimated, perhaps substantially, the study's ability to detect radiation effects, which meant the study results were less definitive than had been reported.[48] The study's weakest link was the HEDR estimated radiation doses from the 1940s and 1950s. The statistical power of the HTDS to detect an association between I-131 and thyroid disease was not as high as claimed due to inadequate allowance for imprecision in the HEDR dose estimates.[49]

HEDR dose estimates were imprecise, in part, because most people were unable to accurately recall how much time they had spent outdoors or in the Columbia River decades earlier or the quantity of each type of food they had consumed at each meal. Additionally, it was impossible for HEDR scientists to determine exactly how Hanford's radiation traveled; where it concentrated; and how much was actually inhaled or ingested or came into contact with a specific individual. Individual characteristics, including age at the time of exposure, sensitivity to radiation (radiosensitivity), amount of thyroid-protective iodized table salt used during childhood, and thyroid gland size at the time of exposure, all impacted the actual I-131 exposure dose.[50] In the words of one nuclear historian, dose reconstruction "is a dicey science because of the extreme patchiness of both human memory and radioactive contamination."[51]

The HIDA Project, sponsored by the Washington State Department of Health, provided individual I-131 thyroid dose estimates for people who were within the HEDR study area between December 26, 1944, when Hanford's off-site radiation releases began, and December 31, 1957. HIDA dose estimates were based on HEDR models, and, like the HEDR doses assigned to HTDS participants and, later, to Hanford bellwether plaintiffs,[52] HIDA doses depended on the best recollection of each person of his or her diet and places of residence decades earlier. At a public meeting, an HIDA project scientist provided a vivid example of the pitfalls of asking people to try to remember minute details of daily life from the past: "This is something we received, the proverbial fruitcake in the diet questionnaire; 'I am 58 years old, and both my parents are deceased. How in the world do you think I could accurately remember the types of food I ate between 5 and 9?'"[53]

Epidemiologist Dr. Joseph L. Lyon, whose study had found increased incidence of thyroid disease and thyroid cancer in children exposed to I-131 from atomic testing at the NTS, looked into why the HTDS findings were so different from the findings of multiple studies of the impact of childhood exposures to I-131.[54] Lyon found that while similar epidemiological techniques had been used to assess the incidence of thyroid disease and cancer in NTS downwinders and Hanford downwinders, the way the dose estimates were calculated for members of the two downwind groups was very different. Lyon hypothesized that this might explain why the negative findings of the HTDS are unique among the studies of children exposed to I-131 from nuclear fallout. Exposure doses from low-dose ionizing radiation downwind of Hanford and the NTS were cumulative, acquired over days, weeks, or years. NTS dose estimates were based on ground contamination measurements taken by off-site radiation monitors following atomic tests.[55] In contrast, HEDR dose estimates were developed decades after exposure, reconstructing the air concentrations and deposition of I-131 at each location where the children had lived, and were based exclusively on model calculations of Hanford releases and historical meteorological data.[56]

The Fred Hutchinson HTDS team attempted to defend the negative findings of the study. They suggested that the increased incidence of thyroid cancer found among NTS downwinders might be explained by examiner bias—meaning that those who conducted NTS downwinders' exams must have known the estimated NTS fallout dose for each person examined and therefore must have performed a more thorough thyroid exam on those with higher doses. This suggestion implied that more thorough examinations detected thyroid cancer, whereas less thorough examinations did not, and that examiners had conducted thorough exams only if they thought the participant had lived in a higher-exposure area. However, there are several reasons why examiner bias could not account for the association found between NTS radiation and subsequent thyroid disease downwind.[57] During exams of members of the NTS downwinder cohort in 1985–1986, the fact that raw milk from backyard cows or goats contained much higher levels of I-131 than processed milk was not yet known and so would not have influenced examiners' beliefs about the exposures of those who had consumed raw milk. Further, 35 percent of NTS downwinders studied had moved since the time of exposure. Examiners did not know where the participants had lived between 1951 and 1958, the years of aboveground atomic testing, and therefore could not have been biased based on knowledge of the participants' possible exposures. Finally, increased incidence of thyroid disease and thyroid cancer was found primarily among those exposed to fallout from shot Harry on May 19, 1953. Examiners did not know which of the study subjects examined had been exposed to shot Harry, so they would not have known who had received the higher exposures from this atomic test.

The Fred Hutchinson team then suggested that the difference might be explained by higher dose rates and a different mixture of radionuclides in the fallout from NTS atomic tests and Hanford radiation releases, a suggestion that Lyon also felt was not plausible. Lyon concluded that the difference was a product of the dependence of the HTDS on mathematical models in order to estimate Hanford downwinders' exposures to I-131. The uncertainties in dose calculations used in the HTDS were substantially greater than those in NTS dose calculations and were larger and more complex than asserted by HTDS scientists.[58] Large, complex uncertainties in dose estimates contributed to the misclassification of doses, impacting the statistical power of the study and the finding of a relationship, or lack of a relationship, between the HEDR estimated I-131 exposure doses and incidence of thyroid disease.[59]

Surveys and Studies That Reflected the Downwinders' Reality

Because the government had lied repeatedly about the safety of Hanford operations, many downwinders deeply distrusted government projects, including the HTDS and HEDR. When the HTDS found no thyroid harm from Hanford radiation releases, that distrust only intensified.

The Northwest Radiation Health Alliance (NWRHA), a downwinders' advocacy group led by Dr. Rudi Nussbaum, a Holocaust survivor and physics professor at Portland State University, conducted a survey of the full range of health issues reported by the downwinders. Less costly and more quickly completed than an epidemiological study, the survey could help counteract the negative findings of the HTDS. The NWRHA survey tracked childhood residence within Hanford's downwind region, period of residence, smoking habits, food sources, diagnosis with any of twenty-nine specific diseases, and information on the health of the participants' children.[60] The survey database included 801 respondents born between

1881 and 1992 who had lived within Hanford's downwind region for at least three months after March 1945. The survey found elevated levels of juvenile hypothyroidism, hyperthyroidism, thyroid cancer, central nervous system cancers and female reproductive cancers.[61] Hypothyroidism in females was thought to be associated with excess reports of spontaneous abortions.[62] Survey respondents also reported lupus, allergies, skin disorders, autoimmune disease, chronic fatigue syndrome, fibromyalgia, multiple chemical sensitivities, depression, and heart disease.[63]

The R-11 survey, funded by plaintiffs' attorney Tom Foulds and endorsed by the Hanford Downwinders' Coalition,[64] looked at the incidence of thyroid cancer and other thyroid disease in Hanford downwinders and found a prevalence of hypothyroidism and hyperthyroidism. Survey respondents reported goiter (swelling of the thyroid gland) and other thyroid disease approximately six to ten times more frequently than respondents to a national health survey.[65] Survey participants' diagnoses were verified by a physician.

The ATSDR, a federal public health agency, investigated the elevated incidence of late-pregnancy issues and neonatal deaths during the years of the highest I-131 releases, 1945 and 1946. The ATSDR reviewed data on 56,320 births, 1,656 infant deaths, and 806 fetal deaths in eight counties surrounding Hanford for the period 1940 through 1950. Pregnant women who lived in areas with the highest estimated I-131 exposures were found to have increased incidence of preterm births. An association with infant mortality was also suggested.[66]

In contrast to the narrow epidemiological focus of the HTDS, these surveys and studies more accurately reflected the broad range of health issues, including autoimmune diseases, cancers, reproductive disorders, neonatal deaths, and preterm births, reported by members of Hanford downwind communities.

Plaintiff 19: Lois Foraker

Lois Foraker, a Hollywood actress, was born in Pasco and moved to western Washington with her family when she was two years old. As a child, she returned every summer to Kennewick. As an adult, after a tumor was found on her thyroid, she underwent a total thyroidectomy. Following the surgery, she developed asthma, also an autoimmune disease. Lois first learned of the possible connection between Hanford *and her health issues in around 1986, when, following the declassification of Hanford records, the media began to report on radiation releases from Hanford.*

This is her story.

I was conceived in Pasco, Washington, and born there on January 1, 1946. Pasco and Kennewick are next to Richland along the Columbia River. Our family moved to the Everett area in western Washington, north of Seattle, two years later. I visited my cousin in Kennewick every summer as a kid to help with her 4-H calf. Pasco and Kennewick were just small towns then; Kennewick couldn't have been more than a few thousand population. Hanford at the time was just that big federal, mysterious, hush-hush project over past Richland that nobody talked about.

I knew from the time I was in grade school that I wanted to be an actress. I was at University of Washington (UW) in Seattle for a year in the acting program. In 1968, I left UW to join the American Conservatory Theater (ACT) in San Francisco.

ACT at the time was one of the premier repertory companies in the United States. I was in the ACT training program for one year and then was cast in various roles over the 1969–1970 season as a member of the company. While in San Francisco, I was also tapped to play featured roles in Clint Eastwood's *Dirty Harry* and *The Candidate* with Robert Redford.

I moved to Los Angeles in 1972 and worked on TV in M*A*S*H. I played various nurses on the show, twice as Nurse Able. I also had the role of Nurse Coleman on *After-MASH*, a show spin-off.

In 1982, I appeared on *Newhart*; in 1984 and 1986 in *Murder, She Wrote*; in *Gremlins* in 1984; *The Exorcist III*, *Child's Play*, *Radio Flyer*, and *Hill Street Blues*, as well as *The Dom DeLuise Show* as Dom's girlfriend. My favorite TV role was that of a Russian housekeeper on *St. Elsewhere*. It was a wonderful role. I speak a little Russian, as I had studied it at UW, so I sounded somewhat authentic.

I was in the pilot for *Northern Exposure* (1990), and I appeared on *Night Court* (1990), *Rachel Gunn, R.N.* (1992), *The Larry Sanders Show* (1995), *3rd Rock from the Sun* (1996), *The X Files* (as Sylvia Jassey) (1999), and *The West Wing* (2001).

I was pretty healthy until I found a lump in my neck in 1982.

I was honestly in terror when I found that tumor. I kept feeling the lump on one side of my neck and checking to see if there was a lump on the

Lois Foraker in 1949. Photo courtesy of Foraker family.

other side. There wasn't. I went in for an exam, and they fed me radioactive iodine. Then I had to be isolated until the radiation left my body. Not fun. Not long after that, I went home for Christmas and broke out in hives all over my body. I assume that it was a reaction to the radioiodine.

When the tests revealed it was time to remove the tumor, the surgeon found that the scoring on the half of my thyroid that didn't have the tumor was so striated that it was highly likely to erupt into tumors, so the surgeon took out the whole thyroid. Just because there weren't growth spores taking the cancer to the rest of my body doesn't mean there weren't problems.

The whole disruption of the endocrine system from having my thyroid out, I think, caused me to develop asthma. I was on prednisone for fifteen years. I had a really bad episode in the 1990s and was intubated for a week. I was doing inhaled steroids, and it was like having thrush; my vocal cords were coated, and I was choking on it all. I have autoimmune thyroiditis and then got asthma, which is also autoimmune. As I understand it, having one autoimmune disease increases susceptibility for other autoimmune diseases.

One of the effects of the thyroid disease was weight gain, which can be a real problem in the world of Hollywood. I was up for one of the lead roles on L.A. Law and was taking steroids. I found that on the steroids, I had too much energy for camera. The steroids cause buzzing. Everything seemed supercharged, like having a turbo engine in your body. I remember, at dinner with a friend, I announced, "I have too much energy in my body!" For film, you are supposed to be very simple, and I couldn't be simple. I lost that role on L.A. Law due to these health issues.

I had also been on 3rd Rock from the Sun, and they called for me to come and do my role. At the time I was in the hospital, intubated, and I couldn't go, as they were shooting me up with five hundred milligrams of prednisone per day to try to get control of the asthma. There was no way I could do the role, so they rewrote my whole neighbor relationship with John Lithgow, and I lost the role.

I also lost my theater relationships, as you can't go onstage unless you have a lot of breath. These health issues altered my entire life. I lost many professional opportunities once I started to get sick.

In many ways, I feel that this life change opened up a new world for me. I am not grateful for the lost opportunities and lost wages, but I began seeing the world as more than just show business. I worked for the last nine years with the city of Beverly Hills as executive assistant to the assistant director of community services. I also served as staff support to the Beverly Hills Fine Arts Commission. I loved that job.

I first learned about the possible connection between Hanford and my health issues from my aunt, who lives in Walla Walla. This was sometime in the mid-1980s, when I was dealing with severe thyroid and asthma problems. Some of my friends in LA had also seen a report on Hanford, describing health problems that sounded like those I was going through. They said, "This sounds like Lois!" I went ahead and contacted Tom Foulds, one of the attorneys mentioned in an article that appeared at the time.

Lois Foraker. Photo courtesy of Foraker family.

Plaintiff 20: Marcy Lawless

Marcy Lawless was born in Spokane and raised in Lind, Washington. Marcy's father had serious health problems and passed away at age sixty of acute leukemia. Marcy believes her weakened immune system, thyroid cancer, gastrointestinal problems, headaches, anxiety, depression, and memory issues may be related to her childhood radiation exposures.

This is her story.

I was born in Spokane, Washington, in 1956. My mother had traveled to Spokane to give birth in a hospital. I grew up in Lind, a small town in Adams County within the Hanford downwind region. I stayed there until I was ten, when we moved to Connell, in Franklin County, another small town even closer to Hanford. We lived there for two or three years and then moved to Spokane, also impacted by fallout from Hanford. I lived in towns downwind of Hanford during my entire childhood.

My dad was born in Walla Walla. He worked in construction in Lind. He suffered from malignant melanoma twice when I was young and then passed away from acute leukemia when he was sixty. My dad always suspected that Hanford was the cause of his health issues. This was way before 1986, when the DOE released all those records. He knew they had built the atomic bomb there at Hanford, so he suspected he had been exposed to something related to Hanford.

All of us kids were born and raised in Lind and Connell. We got our milk from a dairy in the Tri-Cities. Everything we ate and drank came from there. Everything was fresh, not canned.

I have a lot of health issues that I think are related to my childhood exposure to Hanford radiation. I am very sensitive to things—to chemicals, soaps, other substances, and certain drugs like the synthetic thyroid hormone they tried on me because I have autoimmune hypothyroidism. I couldn't handle the synthetic hormone. I finally had to take the natural form.

My immune system sucks. I used to always get sick when our kids were young and brought typical childhood illnesses home from school. I also get sicker than other people. I have had to go to the emergency room twice. The first time, it was because of the flu and the breathing problems I was having, and the other time, just after my thyroidectomy, I couldn't keep anything down.

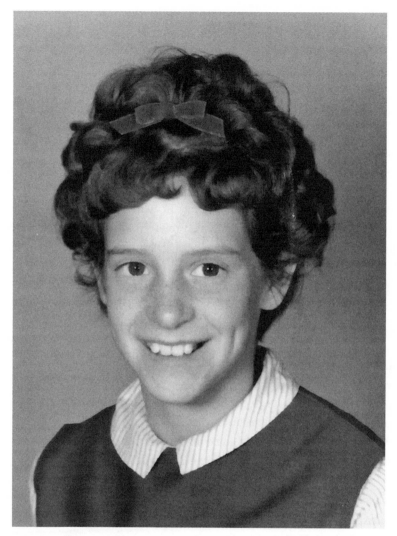

Marcy Lawless as a child. Photo courtesy of Lawless family.

I also have gastrointestinal problems. I'm on Nexium for severe acid reflux. I'm gluten and lactose intolerant. My digestion is crap. Also, I've been constipated for years—I think that's related to my hypothyroidism.

I have frequent headaches, especially behind my left eye. I have acute anxiety disorder and depression, which I have had for many years.

I was having memory issues, so I went to a neurologist. They did an MRI and found lesions like those you see with MS but not enough to be

MS. I used to have a very sharp mind. I worked in the district court for five years. I could recite all the relevant Revised Code of Washington sections from memory, and I routinely told the attorneys what specific forms they were supposed to file. I was in accounting for thirty years before that, and I could remember everything. Now I can remember very little. Some days I can barely recall anything or respond to simple questions. This started happening in 2004, when I stopped working. Doctors said I was in "fight-or-flight" adrenaline-fueled mode all those years. When I stopped working, and the pressure and challenge of work were gone, the adrenaline levels went down, so I wasn't alert anymore. It's possible my adrenals are messed up. But something else is going on as well.

I have autoimmune thyroiditis, also referred to as Hashimoto's. They think my thyroid disease has something to do with the memory issues. I think my brain chemistry is off. I have had panic attacks, and they might be related to periods of hyperthyroidism, which I know can happen with autoimmune thyroiditis. I know that people who are hyperthyroid sometimes have panic attacks. I don't know if I have always been hypothyroid or sometimes hyperthyroid as well. I know that people with autoimmune thyroid disease can be hyperthyroid, then hypothyroid, alternating between the two. I just know my thyroid has been messed up for years.

The doctor I see is a specialist in brain chemistry. It took her five years to figure out the combination of medications that have helped me be a bit more stable. Now I don't go so much up and down—my husband describes it more like a median between the highs and lows.

I had a thyroid goiter for three or four years that kept getting bigger. It got so big that it started interfering with my swallowing. I went to a specialist in Seattle. I wanted the whole thyroid out, but the doctors said that I had very little chance of thyroid cancer, so they only took the half of the thyroid out that had the goiter.

In spite of the reassurances they had given me, they found cancer in the pathology samples they took and had to take the rest of the thyroid out in a second surgery. They said the thyroid cancer they found was stage 3. I'm still fighting to get my full voice back. I have tightness, and it's hard to swallow. My scar is long, as I had a big goiter on the right side that they had to remove along with the thyroid. I also had to have radioiodine treatment, and that was not fun at all.

Marcy Lawless. Photo by author.

I learned about the Hanford radiation releases in 1986, when the DOE released all those records. I filed a personal injury claim in the Hanford downwinder litigation for my thyroid disease, which I feel was caused by childhood radiation exposure. They told me my HEDR radioiodine exposure dose to the thyroid was too low to legally prove that my thyroid disease was caused by Hanford radioiodine, so they took me off the plaintiffs' list. This was about 2011. I got thyroid cancer three years later.

I want people to know the whole story of what my family and the downwinders have gone through. All of us kids in my family are sick. The public doesn't know about any of this. And the fact that the lawsuit went on for so many years is really disturbing. The money the contractors and the government spent fighting our personal injury claims could have gone to us to help us out. The government shouldn't be allowed to drag things out so long, using our tax dollars to fight our own cases.

What happened to us was a long time ago, and people don't think Hanford is a hazard anymore, but it still is. It's the most polluted, toxic site in the country.

The government should have taken care of those it injured. These were honest and hardworking people. I pay the government, through my tax dollars, to protect me. I deserve that protection.

Hashimoto's Thyroiditis

People with Hashimoto's thyroiditis can experience symptoms of hypothyroidism (thyroid hormone too low) or hyperthyroidism (thyroid hormone too high) or sometimes can alternate between symptoms of both conditions.

Both hypothyroidism and hyperthyroidism can cause cognitive problems that mimic the symptoms of dementia. Hypothyroidism can cause mental deficiency and lead to psychosis.[67] Many people with hypothyroidism complain of "brain fog," which is symptomatic of inflammation in the brain. Brain fog can involve problems with both short- and long-term memory and confusion and lead to feeling "spaced out," anxious, or nervous without reason and to feeling tired or unmotivated. Hypothyroidism has been linked to bipolar affective disorders.[68]

Hyperthyroidism can also cause decreased memory and concentration, depression, and other disorders.[69] Those with hyperthyroidism are significantly more likely to report mood disturbances, impaired social function, anxiety, and feelings of isolation and are more likely to be hospitalized with an affective (mood) disorder.[70]

People with Hashimoto's thyroiditis may also experience depression and neurological disorders that resemble or develop into Alzheimer's and Parkinson's disease, including, in very extreme cases, Hashimoto's encephalopathy, which causes drastic changes in the brain that appear very similar to the brain damage caused by Alzheimer's. Autoimmune attacks on portions of the brain, particularly the cerebellum, have also been seen in people with Hashimoto's.[71]

Digestive problems are commonly associated with Hashimoto's thyroiditis. Many people with hypothyroidism do not produce sufficient stomach acid, a condition referred to as hypochlorhydria.[72] Hypothyroidism can lead to delayed gastric emptying: without sufficient stomach acid, food remains in the stomach too long, often causing heartburn and bacterial overgrowth in the GI tract, which can cause bloating and abdominal discomfort.[73] Lack of stomach acid can hamper the body's ability to absorb key nutrients, including vitamin B12, iron, and calcium, which can lead to anemia.[74] People with hypothyroidism also have slowed intestinal motility, ranging from constipation to pseudo- or actual obstruction of the colon. This low motility is often accompanied by abdominal pain and distension and may be misdiagnosed as functional bowel disease.[75]

Hyperthyroidism speeds up digestion, which can result in diarrhea and abdominal cramps. People who have Grave's disease (autoimmune hyperthyroidism) are five times more likely to develop celiac disease,[76] and associations have been shown between Grave's disease and Crohn's, a chronic inflammatory bowel disease.[77]

Plaintiff 21: Jackie Harden

Jackie Harden's family moved around a lot when she was a child. Her father worked first at the Sandia Base in Albuquerque, New Mexico, and then at the Dugway Proving Grounds, a US Army facility where biological and chemical weapons testing took place. When Jackie was in the fourth and fifth grades, the family lived in Kennewick while Jackie's dad worked at Hanford. Jackie remembers the "duck-and-cover" drills in school. She wonders whether living near Hanford, combined with the other places the family lived, was the cause of her family's many health issues.

This is her story.

I was born March 28, 1942, in Terre Haute, Indiana. I spent most of my childhood moving with my family from place to place. When I was about three years old, my father enlisted in the army and was sent to Aberdeen, Maryland, near Baltimore. We went too.

Close to the end of the war, my dad shipped out. When he returned from the service, we moved back to Terre Haute, and he went back to the ordnance depot where he worked before he enlisted.

Then, when I was about seven years old, we moved to Albuquerque, New Mexico, and my dad worked at the Sandia Base there. When we were in Albuquerque, we weren't supposed to tell anyone where my dad worked. I don't know what that was all about. We stayed in Albuquerque about a year and then went back to Terre Haute. When I was in the third grade, we moved to Toelle, Utah. My dad worked at the Dugway Proving Grounds, which was about a hundred miles away from Toelle. We were there about a year.

Then we moved to Kennewick, Washington, a town very close to Hanford. I was there for fourth and fifth grades. My dad worked at Camp Hanford, north of Richland. Then he got a job in Newport, Indiana, where they stored nerve gas, and we moved with him. I often wonder whether he knew what he was doing, raising his children around all these toxic sites. I don't think people understood the risks back then.

I stayed there in Indiana until I graduated from college and completed my master's degree.

There were six kids in my family. I'm the oldest. I have a brother, Michael, who was born in 1947. He was about four years old when we lived in Kennewick. He served in the Vietnam War. Within a year of his return, we learned he had a brain tumor. The doctors thought the brain tumor had probably been there for a long time, but we don't know how long. It could have been a result of exposure to Agent Orange in Vietnam or possibly related to his exposures from Hanford as a child. We will never know. He passed away in 1978 after undergoing several surgeries, cobalt treatment, and chemotherapy.

I was basically a tomboy. In Kennewick, I often played with my brother and his friend. We dug caves in the sand and dirt and in the irrigation ditches when they didn't have water. We made caves and used them as forts. We also played in the flume when it was empty. A flume is where the irrigation water travels, like a railroad trestle. It's a half-circle container with crossbars on top. Water traveled through those flumes across the

Jackie Harden and her brother. Photo courtesy of Harden family.

valley. We played in the dirt and swam in the Columbia River at Sacajawea Park. We got all kinds of exposure from that to Hanford radionuclides in the soil and water, I suppose.

We got our milk from the local dairy in Kennewick. We had an apricot tree in our backyard, so we ate a lot of fresh fruit. I don't remember if we had vegetables. We ate local apples. I guess they were all contaminated with Hanford fallout too.

My younger sister, who was born in Indiana, passed away when she was in her twenties. She had been sick since she was young. She suffered from two truly awful autoimmune disorders. She had both dermatomyositis and scleroderma. Dermatomyositis is a rare inflammatory disease that leads to the destruction of muscle tissue. Scleroderma is a disease of connective tissue that can lead to organ damage. Her calcium levels were also really messed up. She had so much calcium in her body that her elbows, knees, ribs, and vertebrae had fused. She died of pneumonia. The pneumonia wasn't detectable because her fused ribs prevented viewing her lungs by X-ray. She died when she was twenty-three, and by then, she had been ill more than ten years.

I started getting migraine headaches and severe nosebleeds when we lived in Kennewick. The nosebleeds were so bad they scared my teachers. The migraines were truly terrible—when I had a migraine, I had to go into a dark room and try not to move around much for twenty-four hours, and I was deathly ill. I always had to vomit before I felt better. The migraines affected my vision too—I could only see half of everything. Sometimes I had diarrhea as well.

When we were in Kennewick, we went one time to visit one of my dad's coworkers. I remember that they had a daughter about my age. Not long after we visited, we heard that the father in that family was determined by Hanford officials to be a spy, and the family all disappeared shortly thereafter!

In Kennewick, we didn't have fire drills, we had air-raid "duck-and-cover" drills. We would climb under our desks and put our hands over our necks. That protected us from radiation? I never remember a fire drill, always an air-raid drill. Duck-and-cover might protect you from the roof caving in but not from radiation.

I have health issues that I feel were caused by my exposure to Hanford radiation while I was a child in Kennewick. I have Hashimoto's autoimmune thyroiditis. I also have fibromyalgia, and hypopituitarism, causing a deficiency in growth hormone. I have chronic fatigue and dry eyes and mouth. These may be symptoms of Sjogren's syndrome, which is also an autoimmune disease. I have arthritis and had spinal surgery last year. My doctor diagnosed me with mixed connective tissue disorder, polymyositis, an inflammatory disease.

I used to be a workaholic. I served as the assistant dean of the graduate school at Indiana University. Then, I worked on a project for Indiana State University to establish community learning centers in six counties in south-central Indiana. The project took a great deal of energy. I traveled extensively and met with many people. I retired early at age fifty because I no longer had the stamina to work. My fatigue and the pain from the fibromyalgia were just too much.

In addition to being unable to work any longer, I can no longer do so many of the things I used to like to do. I can no longer do much of any yard work, and I can't walk very far. Everything in my life is limited. I have to really pace myself.

I learned about Hanford's radiation releases when I was visiting my daughter in the Portland, Oregon, area. I went to the library one day, and I saw some articles and studies relating to Hanford. From there, I began to learn about the radiation releases that took place while I was a child in Kennewick.

I know some people have serious illnesses and some have even died from Hanford exposures. I joined the lawsuit to add my voice and my story to the litigation. I thought, with more of us as personal injury plaintiffs, we would have more power, and people who were sicker than me could get compensation and help.

The public should know what our government exposed us to and what they exposed workers like my father to as well.

9

In 1990, a government report[1] publicly confirmed what many people had suspected following the mass 1986 declassification of early Hanford environmental monitoring records: Hanford had chronically discharged significant levels of low-level ionizing radiation over a vast area of the Pacific Northwest. After the report was made public, thousands of Hanford downwinders filed personal injury tort claims against former Hanford contractors for cancer and other debilitating illness believed to have been caused by exposure to radioactive fallout released from the chemical separation of plutonium at the nuclear weapons facility.

Remarkably similar arrays of radioactive isotopes were present in fallout released from chemical separations plants at Hanford and in fallout produced by atomic tests detonated at the NTS. The radiogenic cancers and other disease in people downwind of Hanford and the NTS were identical. Yet, as unreasonable as this may seem, the options for relief through the court system available to US civilian downwinders are vastly different, depending upon whether those civilians were exposed to fallout produced by production or by testing of nuclear weapons.

The enactment of the Warner Amendment, which subverted the rights of NTS downwinders to sue private atomic testing contractors, was motivated not by science, logic, or compassion but by a political agenda forced by already indemnified atomic testing contractors who sought redundant protection against personal injury tort suits filed by NTS downwinders.

As discussed in Chapter 7, the Warner Amendment substituted the US government for private testing contractors as the sole defendant in the NTS downwinders' personal injury tort actions. Downwinders' suits pending in state court were removed to federal court, and these suits, now against the government, were adjudicated under the FTCA. The government then shielded itself from liability for the downwinders' cancers and other fallout injuries by asserting sovereign immunity under the discretionary function exception to the FTCA. In *Allen*, on appeal to the Tenth Circuit, application of the discretionary function exception to the actions of the government in its conduct of atomic testing led to the reversal of bellwether verdicts. The Warner Amendment effectively left the NTS downwinders without legal remedy for their injuries.

In 1990, as the first personal injury claims were filed by Hanford downwinders, they found themselves in a far different situation than that facing the NTS downwinders. Hanford downwinders were able to sue former Hanford contractors in federal court through a public liability action, an exclusive federal cause of action created by an amendment to the Price-Anderson Nuclear Industries Indemnity Act (P-AA) in 1988.[2]

Congress enacted the P-AA[3] in 1957 as a comprehensive amendment to the AEA of 1954 in order to make available "a large pool of funds to provide for prompt and orderly compensation of members of the public who incur damages from a nuclear or radiological incident no matter who might be liable."[4] The P-AA defines a nuclear incident as "any occurrence . . . within the United States causing . . . bodily injury, sickness, disease or death . . . arising out of or resulting from the radioactive, toxic, explosive, or other hazardous properties of source, special nuclear, or by-product material."[5] One might ask why the definition of a nuclear incident under the P-AA would not include bodily injury, sickness, disease, or death arising out of or resulting from the radioactive properties of a source or special nuclear material released in fallout from atomic tests.

In February 1991, District Court Judge Alan M. McDonald of the Eastern District of Washington consolidated personal injury claims filed by more than three thousand Hanford downwinder plaintiffs over the preceding year. The consolidated litigation would thereafter be known as *In re Hanford Nuclear Reservation Litigation*.[6] Plaintiffs alleged personal injury, including thyroid cancer, non-neoplastic thyroid diseases, and a range of non-thyroid cancers, as the result of exposure to radioactive and nonradioactive emissions from Hanford.[7] The plaintiffs asserted eight causes of action in tort: negligence and negligence per se, absolute or strict liability, misrepresentation and concealment, outrageous conduct, intentional infliction of emotional distress, and negligent infliction of emotional distress.[8]

The downwinders alleged that Hanford contractors had negligently, recklessly, and intentionally released a vast array of radionuclides and other toxins into the environment. Five contractors that had serially operated Hanford from 1943 until approximately 1987 were originally named as defendants: E. I. du Pont de Nemours and Co., General Electric Co., UNC Nuclear Industries, Inc., Atlantic Richfield Co., and Rockwell International Corporation.[9] DuPont and GE managed Hanford under the AEC prior to the passage of the P-AA in 1957 but were nonetheless indemnified by the US government under the P-AA in the same way as other Hanford contractors, receiving expansive protection with regard to claims relating to the operation of the Hanford facility.[10]

Judge Dismisses Majority of *In re Hanford* Plaintiffs

Judge McDonald divided the litigation into phases. Phase 1 dealt with interrogatory and document discovery. During Phase 2, the generic causation phase, the parties were to focus on generic causation through the preparation of expert reports and expert discovery. Phase 3 was to cover specific (individual) causation.

As discussed in Chapter 7, to prove causation in a toxic tort case, the plaintiff must show both generic causation—that the substance in question is capable, in general, of causing the injury alleged—and specific causation—that exposure to that substance more likely than not caused the specific injury the plaintiff has claimed.[11]

For the Hanford plaintiffs, generic causation referred to whether radiation released from Hanford to the air and to the waters of the Columbia River was capable of causing the cancer or other diseases they asserted.

It could never be directly proven that radiation was the cause of any specific plaintiff's injuries. Instead, to prove generic causation, the plaintiffs would need to rely upon epidemiological studies showing increased incidence of the diseases claimed in populations exposed to similar levels of ionizing radiation.[12]

Specific (individual) causation referred to whether exposure to radioactive fallout from Hanford had caused the plaintiff's specific injuries, taking into account the individual's estimated cumulative exposure dose. In order to prevail, the Hanford bellwether plaintiffs had to prove both generic and individual causation by a preponderance of the evidence.

The plaintiffs and defendants differed in the way they defined the plaintiffs' burden within the "generic" causation phase. The plaintiffs argued that their burden during the generic causation phase was to prove that Hanford's emissions were "capable of causing" their injuries. During the specific causation phase, they would need to prove that they had been exposed to Hanford radiation releases at dose levels "capable of causing" those injuries. The plaintiffs had prepared their scientific experts accordingly, with the expectation that during the latter, specific causation phase of discovery, they would present evidence about the individual, particularized illnesses of each plaintiff and the individual exposures to Hanford radiation. If the requirements of both generic and specific causation were met, the plaintiffs asserted that they were entitled to a jury trial.

The defendants argued that for specific causation, the plaintiffs were required to prove that they had been exposed to a radiation dose at which their risk of the claimed injury was doubled, a level that would statistically double the risk to the plaintiffs when compared to the risk to the general nonexposed population.[13] The defendants asserted that unless the plaintiffs had been exposed to a "doubling dose," an inference could not be made that exposure was more likely than not the cause of the injury, and therefore, the plaintiffs would not be entitled to a jury trial.

During the generic causation phase of discovery, Judge McDonald surprised everyone by dismissing from the litigation, with prejudice,[14] any plaintiff who could not prove by a preponderance of the evidence both generic causation *and* that the plaintiff had been exposed to at least a "doubling dose" of Hanford radiation,[15] the specific causation standard proposed by the defense. Judge McDonald excluded the testimony and evidence of seventeen of the downwinders' proposed expert witnesses, deeming them unreliable or irrelevant if the testimony and the evidence they would present showed only that Hanford's radiation was capable of causing a disease but not that the radiation releases had been sufficient to provide a "doubling dose" to the plaintiffs. The few claims that survived Judge McDonald's summary judgment dismissal were those that met the court's requirements for age at the time of exposure, proximity to Hanford, and doubling dose requirements based on each individual's HEDR dose estimates for thyroid cancer, nonautoimmune clinical and subclinical hypothyroidism, bone cancer, lung cancer, salivary gland cancer, or breast cancer.

The downwinders appealed Judge McDonald's ruling, stressing that based on the judge's own discovery plan, at that stage in the litigation, the parties were still involved in the phase of discovery pertaining only to generic causation. During the generic causation phase, the plaintiffs were required only to prove that Hanford's radiation releases had the *capability* of causing the injuries claimed, not that the specific levels of radiation exposure received by the plaintiffs had caused their specific injuries, which would occur during the specific causation phase. The downwinders argued that they had offered sufficient evidence to meet the burden of proving generic causation and, further, that the doubling dose and doubling of the risk standard had no place in the generic causation discussion because radiation is capable of causing cancer and other disease even at the lowest levels of exposure.

On appeal, the Ninth Circuit sided with the downwinders, reversing Judge McDonald's dismissal of the majority of Hanford plaintiffs and ruling that McDonald had erred in requiring plaintiffs to meet specific dose levels within the generic causation phase of discovery. The Circuit Court found that McDonald had skipped generic causation, determining individual causation without the benefit of full discovery or individual medical evidence.[16]

Furthermore, the Ninth Circuit rejected McDonald's requirement of a threshold doubling dose. That radiation is capable of causing disease, even at the lowest doses, has been recognized by scientific and legal authorities.[17] Earlier Ninth Circuit decisions relied on by McDonald in which doubling doses were required for admissibility of expert testimony involved the morning-sickness drug Bendectin. In the Bendectin decisions, the plaintiffs' case was circumstantial, as there was no definitive evidence showing that the drug had caused the alleged birth defects in the children of mothers who took the drug during pregnancy.[18]

The Ninth Circuit remanded the case to Judge McDonald for resolution of generic causation issues prior to determination of specific causation.

Dr. Thomas Pigford, Special Scientific Master

For the downwinder whose cancer or other disease had been shown through epidemiological studies to be related to exposure to low-dose ionizing radiation, individual reconstructed exposure dose was the key to proving that exposure to Hanford radiation had more likely than not caused the cancer or other disease claimed.

On April 27, 1994, Judge McDonald appointed Dr. Thomas Pigford, a nationally recognized nuclear engineer and retired professor at the University of California, Berkeley, as "special scientific master," a neutral scientific expert agreed to by all parties in the litigation. Pigford's task was to review the validity of the dose estimates of the HEDR study, the first government-sponsored project to estimate radiation doses to the public from a US nuclear weapons production facility. HEDR doses were used for the bellwether trials in 2005 as well as in the HTDS as best estimates of downwinders' individual I-131 radiation exposures.

Shortly after appointing Dr. Pigford, Judge McDonald sealed Pigford's anticipated report, announcing, "Until such time as the Court has reviewed the report . . . the Court intends to keep confidential the contents of the report."[19]

Not long after Pigford delivered his final report to the court in early 1995, its contents were leaked by an unnamed source to the *Spokesman-Review*. The paper brought a motion before Judge McDonald to make the Pigford report public. McDonald denied the motion following a hearing on March 6, 1995.[20]

Judge McDonald reluctantly recused himself from the Hanford downwinder litigation in March 2003 for conflict of interest after downwinders' attorneys discovered that he had purchased an orchard near Ringold in 1999 and had certified to his lending bank that the land was free of radioactive contamination.[21] Critics said that the purchase would influence McDonald's ability to preside fairly over litigation involving whether Hanford's radiation releases had caused harm to communities downwind, as the value of his land and crops could be greatly diminished if it was found that Hanford radiation had resulted in health issues in the area.

Following McDonald's recusal, Judge William Fremming Nielsen took over the litigation. Shortly thereafter, Judge Nielsen asked lawyers for both sides whether there were any grounds for keeping the Pigford report sealed. When none were voiced, Judge Nielsen unsealed the report, making its contents public. Pigford's findings confirmed many of the downwinders' concerns about the HEDR Project. In his report, Pigford concluded that the HEDR Project was flawed and may have underestimated radiation doses to Hanford downwinders.[22] Pigford found several technical issues that increased uncertainty in HEDR estimates of the amount of radiation released from Hanford, which could result in even greater uncertainty in HEDR doses.[23]

In 2005, for the bellwether trials, all parties had agreed to the use of HEDR doses, disregarding higher exposure dose estimates for the bellwethers developed by the plaintiffs' experts. Plaintiffs' attorneys later regretted the decision to use HEDR doses in the trials because a number of the bellwethers' doses, including the HEDR dose for bellwether Shannon Rhodes, were too low to allow these bellwethers to prove that Hanford radiation had caused their injuries.[24]

HEDR doses were also assigned to participants in the HTDS, which determined in 1999 that there was no link between HEDR exposure doses and the prevalence of thyroid cancer and disease downwind.[25] If HEDR doses were underestimated, as Dr. Pigford had concluded, then this problem would have affected the outcome of the HTDS, an anomaly among studies of downwind populations exposed to airborne I-131 in the Marshall Islands, Chernobyl, and the NTS.

Downwinders Argue Plutonium Production an Abnormally Dangerous Activity

In September 2004, anticipating bellwether trials in 2005, downwinders' attorneys filed a motion[26] requesting the court to determine whether the chemical separation of plutonium at Hanford constituted an "abnormally dangerous activity." If so, Hanford contractors could be held strictly liable for the downwinders' cancers and other radiation-related injuries regardless of whether the contractors had exercised the utmost care in the production of plutonium.[27]

Strict liability is a tort liability theory based not on the "fault" of the defendant, but on the premise that under certain circumstances, the defendant should pay for damage or injury caused by its actions regardless of blame. Strict liability is imposed on lawful activities that involve extraordinary risk to others. For strict liability to apply to Hanford contractors, the harm the downwinders asserted had to be one of the possible harms that made Hanford's chemical separations an abnormally dangerous activity.[28] The downwinders asserted that their exposure to Hanford's I-131—released into the air during the chemical separation of plutonium—had resulted in thyroid disease and thyroid cancer, a risk known to the scientific community at the time.[29] The essential question before the court was whether "the risk created [was] so unusual, either because of its magnitude or because of the circumstances surrounding it, as to justify the imposition of strict liability for the harm which results from it, even though it is carried on with all reasonable care."[30]

Judge Nielsen sided with the downwinders on this issue. Hanford contractors may have exercised all reasonable care in the production of plutonium, yet the thyroid disease and thyroid cancer alleged by the downwinders were the possible harms that made plutonium processing an abnormally dangerous activity. The risk of harm to the downwinders was so unusual that Judge Nielsen imposed strict liability for injuries that resulted from that activity.[31] The Hanford contractors appealed, and the Ninth Circuit affirmed Judge Nielsen's imposition of strict liability.

Liability could now be imposed on the contractors without a finding that their conduct fell below a certain standard. In applying strict liability, the court had not determined that the Hanford contractors did anything wrong, only that they had engaged in abnormally dangerous activity and that "in such cases, defendants can spread the risk more equitably allowing those who sustained harm to recover without having to prove that defendants were negligent."[32] *In re Hanford* had now become a causation case: the remaining issue for trial was whether the release of I-131 and other radionuclides from the chemical separation of plutonium at Hanford had caused the injuries asserted by the downwinders.[33]

The Bellwether Trials

In 2005, fifteen years after the first claims had been filed in the Hanford litigation, twelve Hanford downwinder plaintiffs were randomly selected as bellwethers for jury trials. Judge Nielsen hoped that jury decisions in the bellwether trials would encourage settlement in the remainder of the cases.

Six bellwethers did not proceed to trial, either because they withdrew their claims or because Judge Nielsen ruled that, based on the HEDR doses calculated for these bellwethers, they did not have strong enough cases to proceed to trial. The remaining six bellwether cases were tried in April 2005. Bellwethers Gloria Wise and Steve Stanton suffered from thyroid cancer. Wanda Buckner, Shirley Carlisle, and Kathryn Goldbloom were diagnosed with autoimmune hypothyroidism. Shannon Rhodes had lung cancer that her doctors concluded was Hürthle cell thyroid cancer that had metastasized from her thyroid, part of which had been surgically removed.

Because the Hanford contractors had been held by the court to be strictly liable for damage caused by chemical separations at Hanford, the downwinders did not have to prove negligence. As a causation case, the bellwethers' chances of prevailing came down in large part to whether their individual HEDR doses were high enough to more likely than not (a greater than 50 percent chance) have caused their cancers or other disease.

The bellwether trials lasted only four days, followed by four days of jury deliberation. The jury found in favor of Stanton and Wise, both of whom had relatively high HEDR doses from childhood exposure to Hanford I-131. The jury awarded Stanton $227,508 and Wise $317,251 in damages. The jury verdict was a historic first. A jury had never before decided that a US nuclear bomb plant had injured civilians downwind.[34]

The jury reached verdicts against the three bellwethers diagnosed with autoimmune thyroiditis. Roy Haber, one of the downwinders' attorneys, reported that the science wasn't clear on the connection between I-131 exposure and autoimmune thyroiditis but that it had recently become clearer and would be more convincing in the next bellwether trials.[35]

The Tragic Case of Bellwether Shannon Rhodes

The third thyroid cancer bellwether was Shannon Rhodes. Born in Spokane in 1941, Rhodes grew up on a farm near Colfax, Washington, northeast of Hanford. She was diagnosed with thyroid disease in 1978, but her thyroid cancer was not detected until 2002, when she underwent a chest X-ray for a recurring cough. By then, the cancer had metastasized to her lung. Following treatment and a thyroidectomy, the cancer remained in remission, returning years later, with metastatic growths entering her brain and enveloping her trachea. A metastatic tumor in her lung pressed on her trachea, aorta, and superior vena cava.[36] Defense attorneys insisted that the jurors not be told that Rhodes's cancer, which had been in remission, had returned during the bellwether trials.[37]

Based on Rhodes's relatively low HEDR dose, she was unable to prove by a preponderance of the evidence that her exposure to Hanford radiation had been high enough to have caused her thyroid cancer.[38] The jury deadlocked 10–2, one short of the majority Rhodes needed. Judge Nielsen declared a mistrial.

Rhodes requested a retrial, which was expedited because of her terminal cancer diagnosis. On retrial, the jury, unlike the jury in her earlier trial, was told that Rhodes's aggressive thyroid cancer was terminal and that she was expected to die within two years. During jury selection, attorneys for GE and DuPont, the contractors that had operated Hanford during Rhodes's childhood, asked prospective jurors whether they could set aside their sympathy for the dying woman and objectively evaluate whether her childhood radiation exposure, based on her HEDR estimated I-131 dose, was high enough to have caused the cancer that would soon kill her.

Rhodes's retrial in November 2005 lasted two weeks. Rhodes's attorneys asked the jury for a verdict of $20 million to $30 million.[39] At trial, the contractors raised alternative theories about the cause of her thyroid cancer, including X-rays from a tonsillectomy she had undergone as a child.

A DuPont lawyer sat through the bellwether trials and the Rhodes retrial. The proceedings were closely watched by the Department of Justice and DOE contractors, concerned about potential liability from future lawsuits that might be filed by thousands of other people exposed to radiation from nuclear weapons plants.[40]

During Rhodes's retrial, lawyers clashed over the HTDS finding that Hanford had not caused thyroid harm to downwind communities. The jury was not told of a major ongoing controversy over the HEDR doses. Documents obtained by the downwinders revealed that the HEDR had been initiated in part to defend the government against lawsuits by people exposed to Hanford radiation.[41] According to Tom Foulds, an attorney for the downwinders, these documents contained "startling evidence" that the HEDR was designed to "support the litigation positions that the government and Hanford defendants anticipated." That support, according to Foulds, included choosing radiation doses that minimized radiation exposures.[42]

The jury in Rhodes's retrial voted 11–1 in favor of the government. Rhodes hung her head and then hugged her attorney, Dick Eymann. "I'm heartbroken," Rhodes said. "These corporations did this to me and half of Washington State, and now they won't be held accountable. . . . The downwinder issue will probably die, along with a lot of downwinders who will not get to tell their stories."[43]

Rhodes filed a motion for a new trial, alleging jury misconduct. During deliberations on retrial, two jurors had brought up evidence that had not been introduced during the retrial proceedings—that the retrial was Rhodes's second trial and that she had lost the first.[44] Judge Nielsen denied Rhodes's motion, saying that the evidence had not influenced the jury decision.

The downwinders appealed the verdicts against the three non-prevailing bellwethers and the judgment against Rhodes, while the defendants appealed the verdicts entered in favor of the two prevailing bellwethers. The Ninth Circuit affirmed the defense verdict against Rhodes, defense verdicts against three bellwethers, and plaintiffs' verdicts in favor of Stanton and Wise.

Downwinders Face "Scorched-Earth" Defense

At this point, In re Hanford had spanned sixteen years, eleven years longer than it had taken Judge Jenkins to reach his landmark plaintiffs' verdicts in Allen.

For some, the quest for justice through the courts was taking far too long. Plaintiff Harriet Fugitt grew up in Benton City, twenty-six miles southwest of Hanford. Fugitt blamed her disabling health problems on childhood exposure to Hanford radiation. In 2010, nineteen years after signing on as a personal injury plaintiff, Harriet finally gave up hope of ever seeing justice. She and her husband, Warren, took their lives in a murder-suicide, dying from gunshot wounds to the chest. Harriet left a note on her nightstand indicating that the couple's motive was Harriet's Hanford-caused health problems. Police found a twenty-gauge shotgun at the foot of the bed between the couple's feet, two shotgun casings and shells, and an empty bottle of hydrocodone prescribed to Harriet.[45]

In the years following the bellwether trials, with the litigation in its second decade, downwinders' attorney Dick Eymann reported that the plaintiffs would be forced to accept settlement because taking their cases to trial individually at that point would be too expensive. The cost of multiple depositions for each plaintiff and for experts' reports would have to be paid in advance from the dwindling funds available to plaintiffs' counsel, something they could no longer afford.[46] Eymann complained, "I'm not happy at all. From the very beginning we dealt with a scorched-earth defense," referring to the fact that attorneys for the Hanford contractors, fully indemnified under the P-AA, had run up an estimated $80 million in legal expenses, all paid out of tax dollars, to aggressively defend the contractors against the downwinders' personal injury claims. "These people were never fairly compensated for the extent of their injuries and how they suffered from their injuries over many, many years."[47] The downwinders complained that the aggressive defense of the contractors in the litigation, including protracted legal maneuvering over many years, contradicted the mandate of the P-AA to "provide for full and prompt compensation of all valid claims" from nuclear incidents.[48] According to Louise Roselle, lead counsel for the downwinders, "Everything in this case has been unusual, including the fifteen years it took to get to trial for the bellwether cases."[49]

The plaintiffs' attorneys represented the downwinders over the many years of the glacially paced litigation on contingency. The downwinders paid nothing for decades of representation unless there was recovery through jury verdict or settlement, in which case the downwinders' attorneys received the percentage of recovery that had been agreed upon at the start of representation. Legal representation on contingency was critical for many of the plaintiffs who, due to the damaging impact of illness and disability on their income-earning capability, could not otherwise afford legal representation.

By late 2015, many claims for thyroid damage and a few claims for cancers from radiation exposure from the Columbia River had been settled. The settlements were far from generous. Eymann complained, "I thought all [settlements] were very unreasonable given injuries. Some were catastrophic. Even those who were not badly injured were not paid a reasonable compensation for what they went through."[50] Claims had been dismissed for most of the plaintiffs with disease and cancer unrelated to the thyroid and for those with thyroid damage whose HEDR doses were too low to prove causation.

Some of the plaintiffs interviewed for this book told me that they believed that the specific settlement amount they had received was confidential. I looked at the "Settlement and Release of All Claims" form that I had signed when I received my own settlement and found that it made no mention of confidentiality. Nor did I find a reference to confidentiality in a memo from the court-appointed arbitrator regarding the determination of individual awards.[51] I do not recall agreeing to nondisclosure. Nonetheless, Kevin Van Wart, lead attorney for the defense, insisted in interviews with the media that the settlements were confidential.[52] Since the settlements were funded with taxpayer dollars, why were they not public? Perhaps attorneys in the case had stipulated to some sort of confidentiality, but I do not see how this would be binding on the plaintiffs as, to the best of my knowledge, we made no such agreement. Some of the plaintiffs I interviewed seemed confused about the issue, and several told me that they *wanted* the public to know how much the downwinders had received since the settlement sums, in their opinion, were grossly inadequate in light of the degree of suffering the downwinders had endured.

Plaintiff 22: J. M.

J. M. was one of the plaintiffs whose claim, unrelated to thyroid damage, was dismissed from the litigation. She believes that her health and the health of her parents were damaged by exposure to Hanford radiation.

This is her story.

I was born in 1946 in Walla Walla, Washington. We moved to the coast for two years but were back in Walla Walla by mid-1948, where I remained until about 1996.

Our milk was delivered from the local dairy. Our vegetables all came from a huge family garden. I would imagine both milk and vegetables were contaminated by airborne radiation from Hanford. I know that Walla Walla got a lot of fallout from Hanford, especially from the Green Run in 1949.

I learned about Hanford's radiation releases from TV and newspapers soon after the documents were released by the DOE in 1986. I was shocked.

I believe that my health has been damaged by childhood radiation exposures. I have had both basal and squamous cell cancers removed several times. I think that my parents died of cancers caused by Hanford radiation that traveled to Walla Walla. My mother died of leukemia. My dad had stomach pain, and within six weeks, he was dead from pancreatic cancer. We didn't really have time to process such a huge loss, it happened so fast. They were both very aggressive cancers.

I have one sister. She doesn't have health issues that she feels are related to her childhood Hanford exposures. Her husband was a downwinder as well, growing up in Walla Walla. He died of multiple sclerosis.

My husband is also a downwinder. He has Graves' disease, also called hyperthyroidism. He lived as a young adult in Pendleton, then in Walla Walla, both within the Hanford downwind areas.

My husband, my mother, and I signed on as plaintiffs in the Hanford downwinder litigation after my dad died. We also signed my dad on as a plaintiff. Our claims were all eventually dismissed; my husband's Graves' disease, my skin cancers, and my dad's pancreatic cancer. The only claim that survived was my mother's leukemia.

I always worry what may happen next. I worry about my own future due to the deaths of my parents to aggressive cancers.

I remember something very strange. Beginning in 1975, I worked at a local bank in Walla Walla for twenty years. During the time I worked there, there were around one hundred employees in the bank. Virtually every person who worked there had some form of cancer or had a close family member with cancer. We never talked about the possible connection to Hanford, though. It just seems strange to have so many people with cancer.

I want people to be aware of what happened to the Hanford downwinders. None of this was our fault. What happened to all of us was directly due to the action of our own government. We were innocents who got in the way.

Radiation "Hot Spots"

"Hot spot" refers to an area within a radiation contamination zone with higher radiation levels than surrounding areas. Four factors can lead to creation of hot spots: precipitation, wind patterns, stagnation, and impaction.[53]

The weather patterns and the geography of the regions around Hanford, in combination with the nature of I-131, tended to promote the creation of hot spots.[54] Hanford scientists knew about the strong nighttime inversions and distorted flow patterns of the area that could cause these hot spots.[55] I-131 is much heavier than air, and many I-131 plumes moved near the ground, becoming trapped in basins and valleys as higher concentrations, or hot spots, of radioactivity.[56]

The wind generally blew southward from Hanford's emission stacks, toward Walla Walla, and then turned northeast toward Spokane and northern Idaho. Because of the terrain and variations in wind patterns, hot spots of higher radiation sometimes developed. People within these hot spots received higher radiation exposure than those in surrounding areas.

A scientist developing computer models for the HEDR Project believed that the Spokane area might have been a hot spot due to winds that combined plumes of radiation over the area. Walla Walla was also a potential hot spot. HEDR scientists believed that Walla Walla might have become a hot spot during the Green Run, when a combination occurred of calm air on December 2 followed by snow and ice the next day.[57]

Plaintiff 23: Trisha Thompson Pritikin

In January 2013, I was randomly selected as a bellwether for the second set of jury trials, scheduled to begin late in 2013. Our attorneys hoped that verdicts in the bellwether trials would help encourage settlement negotiations.

To prepare for trial, our attorneys asked me to calculate my medical expenses related to thyroid disease, thyroid nodules, hypoparathyroidism, and my total thyroidectomy. I was asked to calculate lost wages resulting from decreased earning capacity related to limitations caused by thyroid and parathyroid disease and related medical problems. Finally, I was asked to describe the suffering I had endured during the decades of untreated severe hypothyroidism that preceded my diagnosis in 1988. That diagnosis was made only after the DOE declassified early Hanford monitoring records in 1986, revealing decades of radiation release, including I-131, from the Hanford facility, revelations that alerted doctors to the possibility that thyroid damage was the cause of my many strange medical issues.

I completed the laborious process of calculating expenses and documenting my health challenges, along with answering a long list of interrogatories from the defense. My deposition was then scheduled in Seattle. During the deposition, I was asked a multitude of questions pertaining to diagnosis with any of a long list of obscure syndromes and diseases that, rather than radiation exposure, might have caused my health issues. The young, auburn-haired associate from the powerful and expensive law firm hired to defend the contractors quizzed me relentlessly, looking for any even remotely plausible alternative causes for the health problems that I firmly believed were due to my childhood exposure to Hanford radiation. To her question regarding diagnosis with any of the syndromes and diseases on her list, none of which I had, I answered, "No," "No," "No"—it became mechanical after the twentieth "No."

She then asked whether I had proof that I had spent time on the Columbia River. This question pertained to radiation exposure I had gotten from swimming in and boating on the Columbia River as a child. I showed her an old family photo of my dad and me, at around six years of age, standing amid the wooden framing of the partially finished boathouse on the Columbia that my dad and his engineering buddies had erected over many weekends in the early 1950s. The structure would house the boat in which we went out on the river nearly every weekend when the weather was even mildly good. My dad proudly flew identical dark-blue triangular "Hanford Navy" flags from both the bow and stern of our outboard as we sped up and down the river. I loved going out on the river to the islands, making mud pies from the wet sand on the island beaches, and swimming in the cold waters. I am not sure why I had thought to bring that photo along with me to the deposition, but it proved to be essential in this instance.

She: "Did you go swimming in the river?"

Me: "Yes, many weekends."

She of the auburn tresses: "How much river water did you swallow?"

Me: "No idea."

She, persisting: "Can you estimate how many gulps of water you swallowed?"

Me: "Uh, not really."

In retrospect, she appeared to be trying to come up with a Columbia River exposure dose from my guesstimating—this was one of the stranger questions I was asked. And on it went.

I didn't much care how much work I had to do to prepare for my upcoming role as a bellwether. All that mattered was that I would finally have the opportunity to tell my story to a jury of my peers. Through my story, the public would come to understand how badly my family and others like us had been deceived by our own government and the harm that that deceit had caused. From my story, the public would clearly see the callousness of the United States toward its own citizens—civilians placed in harm's way.

I desperately wanted to share my story. As a personal injury plaintiff, I had waited almost twenty-four years for this opportunity. As a bellwether, I would finally have the chance.

Then, the bellwethers learned that our cases would not go to trial after all, as all sides had entered into serious settlement negotiations. While final settlement of the cases, if fair, was something the downwinders sought, I was enormously frustrated that I would no longer have the long-awaited opportunity to share the story of our life and loss downwind.

Until now.

This is my story.

I was born October 26, 1950, at the AEC's Kadlec Hospital in Richland. At the time, Kadlec was a closed facility, offering medical services solely to Hanford employees, their families, and others allowed to reside within government-controlled Richland.[58] I grew up in an F house, one of the look-alike alphabet houses of Richland. My father worked at Hanford as a nuclear engineer in the 100 Area, overseeing the operations of Hanford's plutonium production reactors.

My childhood was a happy one, particularly my years at Sacajawea Elementary School, located kitty-corner from my house on Stevens Drive. "S-a-c-a-j-a-w-e-a" was one of the first words I learned to spell. At Sacajawea, I had several teachers who had been transferred by the AEC to Richland from teaching positions at the former Manhattan Project site at Oak Ridge, Tennessee. My parents were amused when I picked up a bit of a Tennessee accent from my teachers.

There were a lot of young kids around my age on our street. We were fond of parading noisily up and down the block on weekends and after school, our mothers standing guard from the doorsteps of our A, B, D, and F houses. These processions featured trikes with multicolored handlebar streamers and, for the older kids, bikes with playing cards attached with clothespins to the spokes, magically re-creating the sounds of a motorcycle.

I remember when one of those neighbor kids, who lived just a block away, was diagnosed with leukemia and started to lose her beautiful red hair that she had always kept in long braids. I was very young at the time and didn't really understand what was happening to her or why everyone was so sad.

My mother and father were very protective of me, their only living child. They had already suffered the loss of my sibling, a brother, who had died shortly after birth. My parents isolated me from my peers for several months as the polio epidemic swept through the Hanford region in the early 1950s, allowing me back into the social scene only after the Salk vaccine became publicly available in 1954. I was routinely hustled into our house whenever the Benton County jeep passed slowly along our street, billowing huge white clouds of faintly sweet-scented DDT behind it to kill the mosquitos that were prevalent in the area. I spent those DDT days inside, my nose plastered against the front picture window of our F house as

Trisha Pritikin and her father in Richland. Photo courtesy of Pritikin family.

I longingly watched my friends frolic through the magical fogbank behind the jeep.

If for no other reason, my parents' insistent protectiveness is how I know in my heart that my mother and father did not sense the danger to which the children in communities around Hanford were being exposed. I am convinced that they had no clue that our air, water, and food, particularly milk and dairy products, were laden with radioactive iodine and other radioactive by-products of the chemical separation of plutonium. If they had had any idea of the danger posed by the facility, particularly the hazards posed to developing children, our small family would not have stayed in Richland. I have come to believe that those in positions of authority at Hanford either denied to lower-echelon employees such as my father that radiation was being released from the facility or blatantly deceived these employees, creating the belief that even if a little radiation was released from time to time, Hanford operations posed no harm to families off-site.

We moved from Richland in late 1960 when my dad was transferred by his employer, GE, to California. Everyone who knew me back then remembers me as a pretty healthy kid, other than having knee deformities and bad allergies. I had had both a tonsillectomy and adenoidectomy by the time we left Richland. A lot of kids were having tonsillectomies back then.

I began to experience strange health problems when I was around fifteen. This is characteristic of "low-dose" exposure to ionizing radiation. With this kind of exposure, years or even decades can pass before the first symptoms of radiation-caused disease appear. This is referred to as a latency period, and the length of latency varies between individuals.

I began gaining weight without relation to the amounts I ate or to my exercise levels. When I was eighteen, my menstruation stopped altogether. At the time, we were living in northern Spain, where my dad, working for an overseas division of GE, oversaw the construction of a nuclear reactor near the small town of Vitoria-Gasteiz. When my periods stopped, my Spanish gynecologist became quite concerned, which caused my parents, in turn, to become equally concerned. I was sent back to the United States for treatment while my parents remained in Spain.

After returning to the United States, I attended Whitman College in Walla Walla for a year and then went back to Europe to be with my family.

I attended the University of Madrid and the American College of Switzerland in Leysin, returning to the United States to attend the University of Washington in 1973. Around this time, I began to suffer from intense, unremitting fatigue. I pushed through and managed to complete my bachelor of science degree in occupational therapy, but my constant fatigue forced me to work limited hours as an occupational therapist once I finished my degree.

I started graduate studies in special education in 1977, as I thought this would require less energy than holding down a job. I soon came down with what seemed to be a very serious case of the flu. Then, a tumor appeared on the side of my neck, just below my right ear. I was very sick, and my lungs were filled with congestion. I coughed almost constantly and was hoarse for several weeks.

The tumor really scared me—I thought it might be cancer—but the biopsy was benign. It turned out I had cat scratch fever, a rare bacterial infection for which people with weakened immune systems are most at risk. I was only in my twenties and had led a healthy lifestyle—why was my immune system compromised? Nobody could tell me.

The "tumor," which turned out to be extreme swelling in my lymph nodes, eventually went away, along with the flu-like symptoms. But my headaches and fatigue got worse, along with increasingly severe digestive issues. And still, doctors couldn't figure out what was wrong.

In 1980, with degrees from the University of Washington in occupational therapy and special education, I wanted to combine my interest in working with people with disabilities with training in law. I had volunteered for several years with the guardian ad litem program in King County as an advocate for special needs children and felt this combination of skills could allow me to do meaningful work, even in a part-time capacity. I was admitted to the University of California Hastings College of Law in San Francisco in 1980.

I continued to suffer from digestive troubles and exhaustion. In 1981, I experienced my first panic attack. I broke out in sweat all over my body, my heart raced, and I was dizzy and light-headed. Then, at the law office where I was interning as a summer law clerk, I suffered what appeared to be a heart attack. With extreme pain emanating through my chest, I collapsed on the floor, folded into the fetal position. I was rushed to the emergency room of a nearby hospital. It turned out it wasn't a heart attack but rather a "cataclysmic contraction" of my esophagus, an exceedingly rare medical crisis that I later learned was triggered by my thyroid gland, adjacent to the esophagus, being highly inflamed from an autoimmune process that was slowly destroying the gland.

Throughout the remainder of law school, I experienced a kind of brain fog and had trouble concentrating. The panic attacks and exhaustion continued. I began to suffer from chronic constipation, and my digestive issues increased. I could keep little food down, and my weight plummeted. Somehow, I managed to graduate in 1983. I am proud of that accomplishment, particularly considering the trying circumstances confronting me.

Painkillers continued to be a constant in my life after law school. A good day was a day with a tolerable headache, where, if I ate carefully, I could make it through the day without feeling too sick. A bad day meant a major migraine, panic attacks, severe indigestion, fatigue, and dizziness. I found myself able to work only limited hours and only some of the time.

None of the medical tests, during countless doctor visits, revealed that the cause of all these mysterious health problems was severe autoimmune thyroiditis.

In 1988, I was visiting my grandmother in Spokane, Washington, and I happened to read an article in the *Spokesman-Review* by Karen Dorn Steele describing airborne radiation that had been released onto unsuspecting communities by the Hanford facility. The article I read described radioactive iodine releases in the late 1940s. I was at first relieved, since I was born in 1950, but would soon learn that those radiation releases continued for several decades thereafter.

I will never forget how angry I was when I first understood that I had been involuntarily exposed to ionizing radiation, one of the most hazardous toxins known to man. I was exposed from the time of conception, in the womb, from the very beginning of my existence! I hadn't had a snowball's chance in hell to escape.

By the time I learned of my childhood radiation exposure in 1988, I had been married three years, and I was in the midst of progressively more intrusive testing in a fertility clinic to determine why I had been unable to become pregnant. I brought the *Spokesman-Review* article about Hanford's radiation releases to the fertility clinic, and they did additional specialized thyroid blood tests, as there was a documented connection between exposure to radioactive iodine, thyroid disease, and infertility.

The testing finally revealed that I was severely hypothyroid and that my TSH level was off the charts. Had the DOE informed the public decades earlier about Hanford's release of I-131 and other radiation, I would have been correctly diagnosed years earlier. This would have prevented years of suffering from undiagnosed or misdiagnosed disease for me and many like me. I honestly do not see any national security justification for the DOE to have hidden the truth from the public for so many years, especially when doing so caused many of us exposed to Hanford's radiation to suffer from worsening, undiagnosed, and sometimes life-threatening health issues.

The fertility clinic immediately referred me to an endocrinologist, who was amazed that I was able to function at all and who voiced concern that with such extreme hypothyroidism, I could fall into a potentially lethal hypothyroid coma. Further testing revealed that I had Hashimoto's autoimmune thyroiditis. My thyroid gland was small and atrophic, indicating that the development of my thyroid had been arrested in childhood as the result of daily exposure to Hanford's I-131.

The endocrinologist explained to me that the panic attacks during law school were likely symptoms of transient hyperthyroidism. People with autoimmune thyroiditis often suffer temporary hyperthyroidism with all the symptoms of Graves' disease and then revert to hypothyroidism.

I went to the medical library at the University of California, San Francisco, near where I lived, to research whether I could pass the effects of my radioiodine exposure on to any children I might have should I somehow manage to become pregnant. At the library, I located the *Life Span Study*, the long-term study of the health effects in survivors of the 1945 atomic bombings of Hiroshima and Nagasaki, which indicated that no genetic effects relating to radioiodine exposure had as yet been found in the children or grandchildren of survivors.

Reassured, I became pregnant the month after I was placed on daily synthetic thyroid hormone. It was years later, after the birth of my second child, that I learned about all the other radionuclides, in addition to the radioiodine, released from Hanford.

Throughout their childhood, my children had a mother who was constantly fatigued, unable to interact with them with the same level of energy they saw in their friends' parents. The thyroid hormone that I must take every day for the rest of my life did not provide the energy and well-being that I had hoped for, although it did improve some of my symptoms.

In 2009, a nodule was found on my thyroid during my annual physical exam, and Hürthle cells, an early indicator of thyroid cancer, were detected in a biopsy of the nodule. My father had died of aggressive thyroid cancer. Like my father, I had been exposed to radioiodine from the Hanford facility. I was likely at the same or higher risk of thyroid cancer as my father because my exposures had occurred during childhood, the most radiation-sensitive period of life. The decision was made to surgically remove my damaged thyroid.

During the surgery, due to the extreme inflammation in my thyroid related to autoimmune thyroiditis, in spite of the best efforts of a very experienced thyroid surgeon and her team, I lost three of four parathyroid glands, tiny glands that sit under the thyroid, regulating blood calcium levels. One parathyroid gland was transplanted under my clavicle by the surgeon in an attempt to cause it to reestablish a blood connection and, thus, functionality. That attempt failed. I now have hypoparathyroidism as the result of the loss of all but one of my parathyroids. Hypoparathyroidism, meaning I have too little parathyroid hormone in my blood, causes tetany that I am not able to fully control. Tetany takes the form of excruciatingly painful muscle contractions and spasms that contort the musculature of my feet and arms, rendering them immobile. Intravenous calcium must be administered immediately to control the tetany. Hypoparathyroidism also increases the risk of kidney failure caused by excessive calcium and phosphorus in the blood.

In addition to the illnesses, injuries, and other issues I have described, I witnessed the rapidly worsening disabilities of my mother and then suffered through her death from malignant melanoma and the death of my father from thyroid cancer. All of this happened to two people I loved, people who trusted the assurances of the Hanford operators that it was safe to live in Richland, next to the Hanford facility, and safe to raise a family there. Both my parents died well before their time. Theirs were painful deaths from cancers recognized as radiogenic and were more likely than not caused by their exposure to Hanford's radiation.

Trisha Pritikin. Photo courtesy of Pritikin family.

The stories of families like mine, America's nuclear guinea pigs, must not be forgotten. Our stories provide incontrovertible evidence of the human toll of nuclear weapons, from production to detonation. May our stories guide us toward a nuclear-free future.

10

The reversal of *Allen*'s bellwether verdicts by the Tenth Circuit Court of Appeals reignited congressional debate over the merits of compensation versus litigation for radiation-exposed populations and other toxic tort victims. Legislators were wary of schemes like the 1969 Black Lung Program, which provided compensation to coal miners who developed fatal degenerative respiratory disease.[1] They worried that, as with the Black Lung Program, the number of eligible claimants could expand uncontrollably, costing taxpayers billions.

The two congressional champions of the NTS downwinders' cause were Sen. Ted Kennedy (D-MA) and Sen. Orrin Hatch (R-UT). The senators felt that the downwinders should receive compensation under a federal program, but they differed on how eligibility should be determined. Senator Hatch believed that radioepidemiological tables should be used to statistically determine whether a person's cancer might have been caused by exposure to fallout from atomic testing.[2] Hatch felt that this objective scientific approach to causation would limit the number of illegitimate claims, thereby addressing legislators' concerns that a downwinder compensation scheme could become another costly entitlement program similar to the Black Lung Program. Senator Kennedy favored "event-specific" compensation, with eligibility based on diagnosis of a specified cancer or other radiogenic disease along with proof of physical presence within areas of fallout contamination during specific time periods.

The Carter administration strongly opposed Kennedy's event-specific approach, concerned that it could become a cost-prohibitive model for other toxic exposure cases pending before the federal courts, including cases involving exposure to Agent Orange, asbestos, benzene products, cotton dust, and beryllium.[3] Hatch, a conservative Republican, was sympathetic to the plight of his downwind constituents but unwilling to adopt Kennedy's event-specific approach. Both the Carter and Reagan administrations were worried that harm to the nuclear industry might result if the information on potential radiation health effects contained within Hatch's radioepidemiological tables was published.

Beginning in 1979, Kennedy and Hatch introduced a number of failed bills to provide compensation to NTS downwinders in geographic areas close to the test site, to uranium miners, and to Utah ranchers who had lost sheep to fallout in 1953. In 1985, following the ruling by Judge Bruce Jenkins in favor of nine *Allen* bellwether plaintiffs and prior to the reversal of those bellwether verdicts in 1987 by the Tenth Circuit, outgoing Utah governor Matheson urged Congress to enact compensation legislation in order to settle the still pending claims of more than a thousand *Allen* downwinder plaintiffs, thereby avoiding the need to try all the remaining cases on a negligence theory. This would allow "[Judge] Jenkins the responsibility of awarding damages to all fallout victims without further legal action."[4]

That same year, in a historic vote on the floor of the Senate, the United States finally acknowledged that atomic testing had caused cancers and deaths downwind—not downwind of the NTS but downwind of tests conducted in the Marshall Islands within the PPG. A portion of a onetime settlement provided by the United States when the Marshall Islands gained independence in 1986 was set aside to establish the Nuclear Claims Tribunal to award compensation to fallout-exposed Marshallese for radiogenic illness.[5] Senator Hatch decided to try a new approach in his continuing efforts to secure compensation for his downwind constituents, introducing an amendment to the Marshall Islands legislation to add compensation for radiation-related cancers and other illness in Utah downwinders. The amendment was tabled. Frustrated, Hatch complained, "Unfortunately, while the United States found it appropriate to deal with this tragedy with respect to people living in the Pacific testing area, it has yet to address the same tragedy with respect to its own citizens equally affected by the Nevada atomic testing."[6]

In 1988, Congress enacted legislation to compensate veterans exposed to atomic tests while in the line of duty. Veterans diagnosed with any designated cancer who established that they had participated in aboveground nuclear tests or the occupation of Hiroshima or Nagasaki after the atomic bombings of August 1945 were presumed to have developed the cancer as a result of their service and were eligible for assistance from the VA.[7] Hatch, persisting in his efforts to obtain compensation for his downwind constituents, observed, "It is clear that if the Federal Government has a duty to compensate the Marshall Islanders and the atomic veterans, it also has a duty to compensate test site downwinders and uranium miners."[8]

Congress Finally Gets the Message

On occasion, disquieting judicial opinions serve as the catalyst for change. The Tenth Circuit's reversal of the *Allen* bellwether verdicts was such an opinion.[9] Many members of Congress were troubled by the reversal of *Allen*, believing this decision represented a grave injustice against the downwinders. The application of the discretionary function exception sovereign immunity defense by the Tenth Circuit to shield the federal government from liability for the downwinders' injuries made it clear that a compensation scheme to bypass litigation was needed in order to avoid future injustice.

In his concurring opinion in *Allen*, Circuit Judge Monroe McKay wrote, "Until Congress amends the discretionary function exception to the FTCA or passes a specific relief bill for individual victims, we have no choice but to leave them uncompensated."[10] It seems that Congress got the message. On October 5, 1990, Congress passed the Radiation Exposure Compensation Act (RECA), introduced by Senator Hatch and signed into law by President George H. W. Bush on October 15.[11]

Hatch finally abandoned his insistence on the use of radioepidemiological tables to establish individual causation. Instead, under RECA, cancers and other diseases were designated as compensable on the basis of the findings of the 1980 Biological Effects of Ionizing Radiation (BEIR) report.[12] RECA is a no-fault compensation system without the requirement that claimants prove causation, one of the nearly insurmountable hurdles facing radiation plaintiffs in toxic tort litigation. Under RECA, when a downwinder claimant meets eligibility requirements, an irrebuttable presumption arises that the claimant developed the eligible cancer or other disease from exposure to fallout from atomic tests at the NTS.

With certain restrictions, downwind claimants who (1) develop primary cancers[13] of the pharynx, small intestine, salivary gland, brain, stomach, urinary bladder, colon, thyroid, pancreas, female breast, male breast, esophagus, bile ducts, liver (except if cirrhosis or hepatitis B is indicated), gallbladder, lung, or ovary or who develop leukemia (except chronic lymphocytic leukemia), multiple myeloma, or lymphoma (other than Hodgkin's disease), and (2) who lived or were physically present in at least one of the "designated affected areas" within Utah, Nevada, and Arizona for at least one or two years between January 21, 1951, and October 31, 1958, depending upon the disease, or who were physically present at any place within the affected area for the entire continuous period beginning June 30, 1962, and ending July 31, 1962,[14] receive a onetime payment of $50,000.[15] On-site participants with designated diseases who were involved in aboveground nuclear weapons tests at US sites and overseas receive $75,000, and uranium miners, millers, and ore transporters with designated diseases who worked in the uranium industry between 1942 and 1971 are eligible for $100,000.[16] No justification is provided in the RECA legislation for the discrepancy in award amounts between claimant categories.

RECA does not compensate downwinders within many of the geographic areas that received significant fallout from NTS atmospheric testing because, at the time that RECA was enacted, the full extent of the dispersion of NTS fallout was not known. The 100,000-page NCI study that detailed NTS fallout dispersion patterns and revealed that NTS fallout had created hot spots of radiation far from the test site was not released until 1997, after many years of delay.[17] The study showed that children living thousands of miles from the test site may have been significantly exposed. Radiation hot spots due to heavy fallout deposition from rainstorms had occurred in areas of Idaho; Montana; the Dakotas; the Midwest; and certain areas of the East Coast, including Troy, New York.[18] The study revealed that children in twenty-four counties, primarily in Montana, Idaho, and Utah, had received the highest exposures from NTS fallout.[19]

Publication of the NCI fallout study prompted congressional efforts to expand RECA to downwinders in additional high-exposure areas.[20] Multiple bills have been introduced by legislators from states that received the highest levels of NTS fallout, hoping to expand RECA to their downwinder constituents. These bills have proposed expansion into regions of Colorado, Idaho, Montana, New Mexico (in particular downwind of the Trinity Test Site), and any county in Arizona, Nevada, or Utah as well as the island of Guam, downwind of testing in the PPG. Several bills have also proposed the equalization of compensation for all claimant categories at $150,000. None of these bills left the various committees for debate on the floor. Undeterred, members of Congress from states that received higher fallout from atomic testing continue to introduce legislation to expand RECA to their constituents.[21]

The only option available to NTS downwinders not eligible for RECA is to file suit in federal court against the US government under the FTCA, as the Warner Amendment prohibits suits against private government testing contractors for injury from atomic testing. As seen by the outcome of *Allen*, the pursuit of justice for test site downwinders under the FTCA has been spectacularly unsuccessful due to the discretionary function exception that continues to shield the federal government from liability for negligence under the doctrine of sovereign immunity.

Medical Monitoring

In 2000, amendments to RECA established a medical monitoring program, the Radiation Exposure Screening and Education Program (RESEP), for NTS downwinders and others eligible under RECA. The program offers physical exams, urinalysis, blood work, chest X-rays as needed, and fecal immunochemical tests. RESEP pays any costs that a patient's primary insurance does not cover.[22] RESEP clinics are located in Arizona, Colorado, New Mexico, Nevada, and Utah.[23]

A medical monitoring program referred to as the Hanford Medical Monitoring Program (HMMP) was recommended for Hanford downwinders by the federal ATSDR.[24] Rep. George Nethercutt (R-WA) and Sen. Patty Murray (D-WA), whose father grew up in Hanford, both endorsed the program.[25] The medical monitoring program was the first assistance the government had offered Hanford downwinders.

The HMMP, which would have screened for thyroid disorders in fourteen thousand downwinders at highest risk from childhood exposure to I-131 released from Hanford, was far more limited in scope than the NTS downwinders' medical monitoring program under RECA. An exposure subregistry also proposed by the ATSDR would have tracked (although not screened for) non-thyroid-related disease and cancers in downwinders participating in the medical monitoring program. In its work to prepare for the start-up of the HMMP, the ATSDR located about seven thousand individuals who had grown up in Benton, Franklin, and Adams Counties, close to Hanford, between 1940 and 1951.[26] This group of Hanford downwinders located by the ATSDR was twice the size of the HTDS cohort. The ATSDR exposure subregistry would have provided a rich source of information about the broad range of diseases in the Hanford downwinder population.

By law, because Hanford is a federal Superfund site, the DOE was obligated to provide funding to the ATSDR for the HMMP and the exposure subregistry. The DOE refused to fund the programs recommended for the downwinders.[27] Many of those of us who had worked for years with the ATSDR to develop these programs felt it was a major injustice for the DOE to refuse to dedicate even a minuscule percentage of its budget to help the downwinders while simultaneously spending billions to attempt to clean up the Hanford site.

Downwinders' attorney Tom Foulds complained, "The true intent of Congress [with Superfund] was to get human remedy as well as physical remedy. . . . DOE has agreed to pay for physical (environmental) cleanup but has refused to provide funds for human (medical) cleanup."[28] Lynn Stembridge, director of HEAL, argued that the DOE's refusal to fund medical monitoring defeated the primary purpose of Superfund, which is to protect communities at risk from toxic exposures:

> One of the many ironies about this [funding] impasse is that the main purpose of the Superfund law is to protect communities, whether it be

from the health risks of past toxic exposures or the potential future risk from toxic exposures. Here's a community we know is at risk today because of past exposures. The cost of the program that would finally begin to help them is literally one percent of the annual Hanford cleanup budget. And the Energy Department is saying this help is not a priority with them. How can this be?[29]

About a year after the public release of the NCI fallout study in 1997, the Institute of Medicine (IOM) recommended against nationwide government-funded thyroid screening for Americans exposed to NTS fallout, gratuitously advising as well against the ATSDR's recommended targeted thyroid screening for those of us determined to be at highest risk from childhood I-131 exposure downwind of Hanford. No justification was provided by the IOM for its recommendation against screening this narrowly defined group of Hanford downwinders.[30] When the HTDS concluded in 1999 that Hanford had not caused thyroid harm to downwind communities, the ATSDR responded by scaling back the HMMP from a medical monitoring program to an "informed decision making" and information program.

As the funding stalemate between the ATSDR and DOE continued, the medical monitoring program for Hanford downwinders was dying before the downwinders' eyes. What was already a narrow thyroid screening program was now reduced to "informed decision making." I was truly fed up with the DOE. The DOE and its predecessor, the AEC, had put us in harm's way as children and young adults, and now the DOE was again turning its back on us. I wanted Hanford downwinders with little or no access to medical care to receive proper screening before a thyroid nodule, like the unmonitored nodule on my dad's thyroid, had the chance to explode into lethal, metastatic thyroid cancer, as it had in my dad's case. Was that too much to ask of the federal agency that had destroyed our health?

Represented by Tom Foulds, I filed a citizen's suit seeking a declaration under Superfund[31] that the DOE was required to fund the HMMP. The district court granted the summary judgment motion filed against me by the DOE, asserting that the court lacked subject matter jurisdiction because I did not meet all the requirements under Superfund for bringing a citizen's suit and because there was no final agency decision that could be appealed.[32] I appealed, and the Ninth Circuit affirmed the lower-court ruling that I lacked standing. I appealed again, this time to the US Supreme Court, which refused to review the case.[33]

The HMMP was officially dead.

Inequitable Treatment: Workers vs. Civilians

Radiation exposure received on the job by nuclear workers was, for the most part, monitored carefully by the AEC. Exposure records for many nuclear workers are retained at the workers' place of employment, and these exposure records are used to help reconstruct cumulative work exposures unless the records can no longer be located due to the length of time since the worker left his or her employment.

The AEC failed to monitor civilian downwinders' radiation exposures, and for this reason downwinders' exposure dose must be retrospectively reconstructed, based in part on individuals' best recollection of residence and diet from decades earlier. For both nuclear workers and downwinders, reconstructed exposures are presented as a range of possible doses within which, for example, with 90 percent or 95 percent certainty, the true exposure dose lies.

Under the Energy Employees Occupational Illness Compensation Program Act (EEOICPA), nuclear workers who develop radiogenic cancer or other radiogenic illness receive the benefit of the doubt throughout the claims process. Nuclear workers at Hanford and a number of other nuclear sites who are diagnosed with any of twenty-two cancers specified under the EEOICPA as radiation related have been designated part of a "special exposure cohort" (SEC).[34] As members of an SEC, these workers and former workers are not required to have their cumulative exposure dose reconstructed and are not required to have the probability that their cancer or other recognized radiogenic disease was caused by on-the-job exposure calculated when they file a claim for radiation injury. The SEC was created to make it easier for nuclear workers to prevail when filing claims for EEOICPA-specified radiogenic cancers and other specified illness and to make up for poorly monitored doses or incomplete worker exposure records.[35]

Hanford workers diagnosed with cancers not designated as radiogenic cancers under the EEOICPA must go through dose reconstruction and determination of the probability of causation to prove their claim.[36] In these cases, the EEOICPA favors nuclear workers in the claims process when determining probability of causation, defining workers' estimated dose as occurring at the "upper 99th percent credibility limit."[37] That is, the estimated exposure dose and the resulting estimate of the probability of causation are defined based upon the dose calculated at the upper 99th percentile of the estimated dose range for that worker. That dose will be used to determine whether the nuclear worker's cancer was "at least as likely as not" (defined as greater than or equal to 50 percent probability) caused by on-the-job exposure to ionizing radiation. This is a more lenient standard than the "more likely than not" (greater than 50 percent) causation requirements imposed on the downwinders in toxic tort litigation.

In the litigation, the Hanford bellwethers' HEDR doses were defined as a single dose number measured in rads, the arithmetic mean of their HEDR interval dose range, rather than as a range of possible doses. If the dose reconstruction and causation requirements applied under the EEOICPA to nuclear workers who are not within an SEC had been applied for *In re Hanford* plaintiffs, each Hanford downwinder's reconstructed dose and probability of causation would have been expressed as a probability distribution reflecting the uncertainty of the dose estimate, and eligibility for compensation would have been determined at the upper 99th percentile of the estimate of the probability of causation. This would have resulted in a far more equitable outcome for Hanford bellwether plaintiffs whose HEDR doses were too low to prove that Hanford exposures more likely than not caused thyroid disease or thyroid cancer.

Attorneys for the downwinders argued that the *range* of possible exposure doses for each bellwether plaintiff should be used to establish causation during the bellwether trials. One of the plaintiffs' experts in the trial of bellwether Shannon Rhodes objected when requested by the court to provide a single dose value as a "best estimate" of Rhodes's true exposure dose but eventually was required to provide the arithmetic mean value of Rhodes's interval range. The defense in Rhodes's retrial argued that the use of dose intervals rather than the arithmetic mean would confuse the jury, who would mistakenly believe that Rhodes's true exposure dose was near the upper boundary of the interval.[38] I argue that it is just as likely that

Rhodes's true dose was found at the upper limit of the range of estimated doses that (with 90 percent certainty) contained her true dose as it is that her true dose was exactly equal to the arithmetic mean of that range.

During the production and testing of atomic weapons, the United States knowingly put its own citizens, including thousands of infants and children, in harm's way. The ethical response to the cries for help from those infants and children, now grown and suffering (and sometimes dying) from cancer and other radiogenic disease, would be for the US government to provide an official apology, along with compensation for radiogenic injury, medical monitoring, and care. Instead, radiation-injured civilians who are ineligible to apply for compensation under RECA have been forced to seek redress through the courts, where the government has aggressively defended itself against their personal injury claims.

Only a small percentage of US civilian downwinders have received compensation and medical care for radiogenic cancers and other disease more likely than not the result of childhood exposure to fallout from US nuclear weapons production and testing. Hanford downwinders are not currently eligible for compensation or medical monitoring under RECA. In bellwether jury trials, just two of twelve bellwethers exposed as children downwind of the Hanford facility were awarded compensatory damages for radiogenic cancer. Of the nearly five thousand remaining Hanford downwinder plaintiffs, only a percentage of those with thyroid cancer, thyroid nodules, and thyroid disease, along with a small number of individuals with river-exposure and other cancers, eventually received settlements from former Hanford contractors. These settlements were insufficient to cover medical bills, pain and suffering, loss of income-earning capacity, and other damages. Downwinders who accepted settlement were required to waive any right to sue the Hanford contractors should additional radiogenic cancers appear in the future. In the words of a medical anthropologist, "Settlements . . . are responses that include no apologies, no concession of wrongdoing, no expression of atonement."[39] Settlement allowed Hanford contractors to close the door on the suffering of the downwinders without admitting that they had caused the downwinders harm.

The disparate response of the US government to nuclear workers and civilian downwinders requesting medical care and compassionate assistance following diagnosis with radiogenic cancers and other disease represents significant injustice against civilians caught in the fallout from nuclear weapons production and testing activities. Civilians' unwitting exposure to fallout from US nuclear weapons production and testing often occurred while these individuals were in utero, infancy, and childhood, the periods of highest vulnerability to the effects of radiation. Unlike adults who entered into employment contracts at nuclear weapons production and testing sites with at least a basic understanding of the exposure risks they would face on the job, civilian downwinders were largely unaware of the health risk posed by radiation exposure. Hanford downwinders were kept completely in the dark over Hanford's radiation releases.

I argue that civilian downwinders, whose exposures were not monitored by the government and for whom no exposure records therefore exist, should be designated, as are many nuclear workers, as part of an SEC, eligible without dose reconstruction and determination of probability of causation, for compensation and health care for cancers and other disease recognized as radiogenic under the EEOICPA for nuclear workers.

Plaintiff 24: Judith Mayer

Judy Mayer is a Hanford downwinder who suffers from Waldenstrom's macroglobulinemia (WM), a form of cancer that is as complex as its name. WM is a type of non-Hodgkin's lymphoma that the government has identified under federal law as related to exposure to low-dose ionizing radiation. Nuclear workers who develop WM at least five years after their first on-the-job exposure receive compensation and health care from the government through the EEOICPA.[40] NTS downwinders who develop lymphoma other than Hodgkin's disease at least five years after first exposure to fallout may qualify for compensation under RECA.[41] Hanford downwinders such as Judy, diagnosed with lymphoma, are not currently eligible for compensation under any government program compensating radiation exposure victims.

Here is Judy's story.

I was born in 1950 in Salt Lake City. My father taught math at the University of Utah. My family moved to Richland, Washington, when I was eighteen months old, and my father went to work for General Electric as a mathematician in its education department.

We lived in a "prefab" house in Richland on Davenport Street. Some of my earliest memories are of playing in the backyards of our house and my

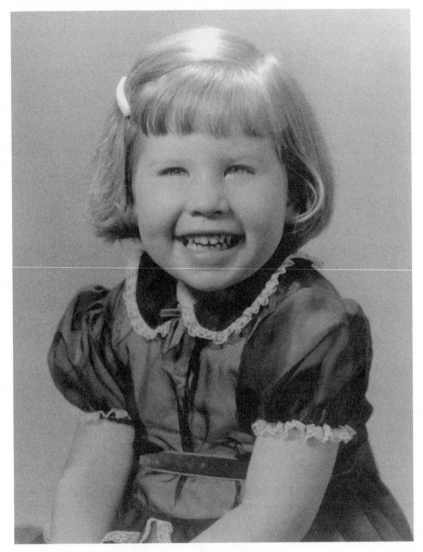

Judith Mayer as a child. Photo courtesy of Mayer family.

friend's house. Fairly often, a siren would go off, and our mothers would hurry us inside until the siren stopped. I also remember drills in school: when the siren sounded, we were instructed to duck and cover under our school desks, as if that were protection from anything.

One summer in Richland, as we were playing in the yard, white flakes

Radioactive Summer Snow in Richland

The summer snow that Mayer recalls may have been "large, flaky particles"[42] of ruthenium from the Hanford facility. Beginning in January 1954, during Mayer's childhood in Richland, Hanford's RE-DOX plant released significant amounts of radioactive ruthenium-106-contaminated ammonium nitrate flakes into the atmosphere at least nine times. Particles and flakes were found throughout the Richland, Kennewick, and Pasco areas.[43] Ruthenium-106 is a beta emitter with a one-year half-life. Scientists were worried that the flakes would cause tissue damage if they came into contact with human skin or cancer risk if inhaled or ingested. Ruthenium-106 particles were found in Richland and on farms across the Columbia from Hanford.

Color maps of the contamination areas were sent to Washington, DC, but no public health warning was issued. Hanford officials considered and then rejected the possibility of issuing a warning; Hanford health physicist Herbert Parker determined that there could be significant possible localized lung doses from the particles. He wrote that he would have liked to check the intestines of local cattle for ruthenium ingestion but "could not do so without the risk of exciting too much comment."[44]

began to fall from the sky. They were like thin shavings of soap. We danced and played in them, pretending it was snow. When we announced to our mothers that it was "snowing" in the summer, they hurriedly rushed us into the house. It's hard to know whether our parents understood the danger from events like this. I suspect they knew that the Hanford project had created nuclear weapons in the war, but Hanford's purpose was never discussed in our family. It was enough to know that Hanford was a major military project important to the country. My dad was a World War II vet, and my parents had great loyalty to the US government, an entity that they had always believed had the best interests of its citizens in mind.

During my childhood, all the kids drank a lot of milk. Mothers believed milk was the best thing to give children to build strong bones. We would learn, decades later, that that milk contained radioactive iodine from Hanford.

When I was five, we moved to a new housing development in Kenne-wick with a beautiful view of the Columbia. In Kennewick, we were envel-oped in abundant agricultural yields. Each summer, farmers from Yakima and Walla Walla drove through our neighborhood selling their crops. My mother would buy a crate of apricots or peaches and then spend her sum-mer canning so that we could eat these fruits throughout the winter. My favorites were the trucks that came through in the fall with the delicious apples. Those crisp red apples would sparkle with a special tartness in your mouth.

The fields were numerous around our development, with miles and miles of grapes to be fermented into wine, a cherry orchard just down the road, and a large field of mint about a mile toward town. When the mint was in bloom, it gave off a wonderful aroma.

In 1960, GE closed the education department where my father worked. He was offered a position with IBM, so we moved to Seattle. We had lived in the Hanford area for ten years.

I later went to the University of Southern California and went on to hold senior positions in three Fortune 500 companies, eventually leaving to found my own consulting business. This required a great deal of travel.

In 1990, when I was forty years old, I came down with what seemed like a cold that wouldn't go away. I was treated, relapsed, treated again, then relapsed. Finally, my doctor decided that something was seriously wrong. She called me home from a trip and proceeded to order every can-cer screening test available. Eventually, she discovered a "monoclonal gammopathy" in my blood and referred me to an oncologist for further testing.

Blood tests revealed Waldenstrom's macroglobulinemia, named after the Swedish researcher who first identified the disease. A painful bone-marrow aspiration confirmed Waldenstrom's, a form of non-Hodgkin's lymphoma, a very rare cancer of the bone marrow.

The doctor told me that the median survival with Waldenstrom's was only five years and that if I was lucky, I would survive ten years. He also told me that I didn't need treatment yet, but once I did, it would be che-motherapy, and I would not be able to stop the chemo once I started it.

Prior to receiving this devastating news, I thought I had more of my life ahead of me than behind me. At forty, I was at the height of my career. I was dazed, in disbelief at what I had heard. That day, as if to defy the reality I was facing, I had dressed in my most sophisticated business suit, looking the perfect picture of good health. I felt fine. This diagnosis could not be true. And yet I had just been dealt a death sentence.

How does one live permanently on chemotherapy? Chemotherapy assaults the body. How long can a body survive this? I felt as if a heavy door was closing on me as I struggled to push it back.

The leading world expert for Waldenstrom's was at the Mayo Clinic. I contacted him for a consultation. He was about to leave the country for a conference but was able to review my records before travel. Over the phone, he confirmed the diagnosis and recommendations of the oncologist but said that my hemoglobin was declining at a steady rate and that I would probably need treatment sooner rather than later.

My husband and I began researching Waldenstrom's. We went to a medical library but found only a one-paragraph description of Waldenstrom's in a medical text. I now understood why doctors would sometimes tell me, "Oh, yes, we talked about this once in medical school," and why nurses always asked me to spell the name of the disease for their records.

The only organization that included Waldenstrom's in its registry was the National Organization for Rare Disorders. Most people diagnosed with this disease are men in their sixties and seventies. So how did a female, age forty, develop Waldenstrom's? No one in my family on either side had been diagnosed with this disease.

I soon found out. My sister-in-law sent me an article from the *New York Times* about the radiation releases from Hanford, revealed in records declassified by the DOE in 1986. The article detailed radiation discharged from Hanford over a long period of time, including the ten years I lived in the area as a child. Since I did not fit the profile for Waldenstrom's, I couldn't help drawing the conclusion that perhaps this illness was related to my childhood exposure from Hanford.

It was from the *New York Times* article that I first read about HEAL. I subscribed to HEAL's newsletters and learned of the dose reconstruction study under way, called the Hanford Environmental Dose Reconstruction Project.

In the meantime, I continued my research on Waldenstrom's and learned that it can be an indolent disease, meaning I could go for years before needing treatment. This information was a huge relief and an extreme contradiction of the initial information I had been given about the median survival of five years.

I was self-employed at the time, and my only medical coverage was as a dependent on my husband's plan. What would this treatment cost? What if my husband lost his job and medical coverage? With Waldenstrom's as a preexisting condition, if I had to change medical insurance, securing a plan that would cover treatment for this disease would be unlikely.

I was advised by an attorney that I should look into the possibility of filing suit against Hanford contractors for personal injury. Proving causation would be the biggest hurdle. There were studies on the health effects of exposure to radioactive iodine, one of the radioactive substances released from Hanford, and the link between exposure to radioactive iodine and the development of thyroid cancer and thyroid disease, but nothing on my disease or its potential cause. Compensation wouldn't cure my cancer. My motivation in considering legal action was primarily to recover the high cost of treatment.

Through reading HEAL newsletters, I followed the progress of the HEDR Project, and when I visited my parents in Spokane, I read the articles written by Karen Dorn Steele in the *Spokesman-Review*. Steele's articles confirmed the historic releases of radiation from Hanford and described Hanford's leaking nuclear waste storage tanks and the seepage of contamination into the aquifer. The contrast between the agriculturally idyllic place where I had lived as a child and the appalling truth that the area's water, milk, and food supplies were dangerously contaminated with radiation was unthinkable. The contamination from Hanford was at best careless and at worst intentional and carried out with callous disregard for the health or safety of the public.

I embarked on a search for studies on the health effects of radiation exposure. The only study I could find was the *Life Span Study* of atomic bomb survivors in Hiroshima and Nagasaki. The information I found was quite dated and not very thorough. Eventually, I came across an article about Dr. Alice Stewart, who was conducting epidemiological studies on the health effects in people exposed to radiation from the Three Mile Island nuclear accident in 1979. I wrote to her, expecting no response, and was surprised when she wrote back. She told me she had not found anyone in her studies with my specific disease. I explored other sources, but my search resulted in a dead end.

In the meantime, the lawyer who had earlier advised me that I had a case for recovery that could help defray the costs of medical treatment indicated he was no longer interested in pursuing my case. I learned from the HEAL newsletter that attorney Tom Foulds had opened the Hanford Litigation Office in Seattle and was taking on Hanford downwind personal injury clients. His office read my file, and they indicated that my chances of financial recovery were slim if not nonexistent. Again, the issue came down to the difficulty of proving that my childhood radiation exposure had caused Waldenstrom's.

I knew at this point that due to my health issues, I could not continue to work in a high-stress job. I returned to graduate school and then embarked on a new career in teaching. This new career did not require travel and offered a salary of only 20 percent of what I had earned before. But, unlike my corporate work, this career was meaningful and gave me purpose.

Then, in 2005, I began to experience light-headedness and extreme itching all over my body. I went in for a blood test. My doctor thought the test results were so concerning that he called me at home on a Friday night and informed me in no uncertain terms that I needed a blood transfusion immediately and treatment as soon as possible. I was panicked.

We discussed treatment options. There were three courses of treatment available, no one better than the other. Because Waldenstrom's was so rare, available treatments consisted only of those that had been successful for other diseases or had had some success in other cases of Waldenstrom's. Nothing was definitive enough for a doctor to say, "This is the treatment you need."

My husband and I began researching once again. Now that some years had passed since my initial diagnosis, there was significantly more information available about Waldenstrom's. The International Waldenstrom's Macroglobulinemia Foundation (IWMF) had been formed, and there was a doctor in a specialized lab at Dana Farber in Boston, the only such lab in the world, whose research was specifically dedicated to Waldenstrom's. We contacted him and flew to Boston for a consult.

The doctor at Dana Farber confirmed that I was in serious trouble and needed to undergo a treatment called R/CHOP [Rituxan/CHOP]. CHOP is an acronym for four traditional, heavy-duty chemotherapy drugs. I underwent this treatment regimen for six months, every six weeks. This involved a six- to seven-hour IV infusion each time. Three difficult weeks followed each infusion. After six months, we saw some success in that the progression of the disease had been arrested, though not eliminated.

For two years thereafter, I continued to receive IV infusions of Rituxan at six-week intervals. Because my body reacted to the Rituxan, the infusion had to be undertaken very slowly over a six-hour period each time. Rituxan does not have the same toxic effect on the body as the other chemotherapy drugs. The discomfort is being hooked up with the IV needle for the infusion and sitting for six hours while the drug flows into the bloodstream.

I was finally able to go off the Rituxan maintenance program. After being off the program for more than a year, my numbers started to slowly creep upward and downward in the wrong directions. Waldenstrom's causes an overproduction of IgM [immunoglobulin M] in the bloodstream and reduces hemoglobin, eventually creating extreme anemia. I needed treatment again, so I opted for Rituxan once more.

After several months, it was clear that the Rituxan was having no effect, and the treatment was discontinued. For a while, I tried plasmapheresis, but this was purely a stopgap measure with no effect on disease progression. Eventually, I started a regimen with everolimus. It was very effective for several years in reducing my IgM. I took it until my numbers started to go south again. At that point, it was the same quandary of which treatment to try, knowing that none of the treatments would cure the disease. Treatment would only put the disease in check for a while, followed by another relapse and further untold damage to my body.

We discovered on the IWMF online listserv that a new drug, ibrutinib, then in clinical trials, was achieving good results in Waldenstrom's patients. This drug provided a targeted treatment that disrupted one pathway of cell division. I called my doctor to see if I qualified for the trial, and he indicated that I might. I contacted the doctor at Dana Farber who was in charge of the trial, flew to Boston, was tested, and was admitted to the trial.

I completed the three-year trial with continued excellent results. During the trial, the ibrutinib was free, but its retail prescription cost is astronomical: $10,000 a month. I'm on Medicare, but my copay for this drug still costs me $10,000 each year.

On top of the Waldenstrom's, in 2006, I was shocked to receive a second cancer diagnosis. At a routine doctor's appointment, my doctor discovered that my thyroid was enlarged. She indicated that this was common but that I should have an ultrasound, just to check it out. The ultrasound detected a suspicious thyroid nodule, and a fine-needle biopsy of the nodule was ordered. The biopsy showed that the nodule was malignant. I was sent to a surgeon who recommended complete removal of the thyroid.

Unprepared to have a crucial body organ removed, I objected, wanting just the malignant nodule removed. The surgeon explained that any tissue he removed would go to the pathology lab. If it came back malignant, he would have to remove the entire thyroid. Reluctantly, I agreed to complete removal of the thyroid. And, indeed, the pathology report confirmed thyroid cancer. Since then, I have taken synthetic thyroid hormone, Synthroid, daily, which I must take for the rest of my life. Several years after starting the synthetic thyroid hormone, I developed low bone density, called osteopenia, as a side effect of Synthroid.

Upon my diagnosis with thyroid cancer in 2006, I was made part of the thyroid cancer group in the Hanford downwinder litigation, and I therefore was one of very few victims who received a financial settlement. The irony is that the disease for which I should have been compensated, the Waldenstrom's, was not addressed in the litigation because no studies had been done on exposed populations to look at the link between low-dose radiation and my specific cancer. This was the case with other cancers suffered by Hanford downwinders as well.

My childhood exposure to Hanford's radiation releases has had a devastating impact on my life. In my opinion, the actions of the Hanford operators were criminal. The initial crime of Hanford was the careless or willful mishandling of radioactive materials during a period in which the country was at war and then the continuation of that operational mentality during the Cold War era. The contractors were indemnified by the US government, so Hanford contractors had no incentive to operate safely.

Where was the oversight? Many people downwind and downriver were put at risk. With the exception of thyroid cancer victims, very little, if any,

Judith Mayer as an adult. Photo courtesy of Mayer family.

compensation has been made available for the victims of the birth defects, the multiple cancers, and the deaths.

I remember the conversations my mother had with her friends, many years after we left the Hanford area, about who had developed cancer and who had died. Many people who lived in our neighborhoods in Richland and Kennewick were affected.

If it had not been for the legal skill and perseverance of our attorney, Tom Foulds, not even the thyroid cancer victims would have received settlements. It took Tom Foulds more than twenty-three years to reach a settlement for the thyroid victims. He was stalled, rulings were made against him by a corrupt judge—delay after delay after delay. I suspect the government paid more money to the defense attorneys than they paid out to victims.

I am lucky. Both of my cancers were diagnosed early. My thyroid was removed before the cancer had spread. Being monitored for Waldenstrom's enabled my doctors to immediately know when I needed further treatment, avoiding a long diagnostic process to determine the cause of new symptoms. And my disease was indeed indolent for many years. I learned not to accept the word of just one doctor but to consult with many specialists; to keep abreast of the research happening all over the world; and to seek out the best, most advanced treatments possible. Those treatments weren't always available to me, and some of the treatment regimens were challenging, but now I am participating in the forefront of cancer therapy, and my current treatment is making a huge difference for me. I now have normal hemoglobin and live without anemia.

Lying in bed, recovering from the painful bone-marrow aspiration that first confirmed my Waldenstrom's, I had a clarifying moment. I realized that I could live as if I were dying or I could live as if I were living. I chose to live as if living. I got up the next day, got dressed, got on a plane, and went on with my work and my life. That is how I have lived since that day. I am grateful for having those years. And that is how I plan to live my future years, however many there may be.

What about the Others?

Hanford, the NTS, and the PPG are not the only US Manhattan Project and Cold War nuclear weapons production and testing facilities that released airborne ionizing radiation off-site. A fallout dispersion study by the NCI to reconstruct doses to civilians downwind of the Trinity Test Site is nearly complete at the time of this writing.[45] If RECA is not amended to include communities exposed to fallout downwind from the Trinity Test Site, then Trinity downwinders will have no other recourse than to sue the government in federal court under the FTCA due to the restrictions facing atomic testing radiation exposure victims under the Warner Amendment. In court, Trinity downwinders will face the very great possibility that any recovery against the government granted by a federal district court judge will be overturned on appeal on the grounds of sovereign immunity under the discretionary function exception to the FTCA.

What about communities downwind of the Savannah River Site, where I-131, tritium, and argon-41 (Ar-41) were released to the air? What about people downwind of the Idaho National Laboratory, which released cesium-137 (Cs-137), phosphorus-32, strontium-90 (Sr-90), and I-131 to surface water and I-131, Cs-137, and Sr-90 to the air? What about communities downwind of Los Alamos, where plutonium and uranium, explosives, and volatile chemicals were released to the air and plutonium was discharged in liquid effluents? What about individuals downwind of the former Rocky Flats facility near Boulder, Colorado, where chemicals and radioactive materials, including americium-241; plutonium-238, -239, -240, -241, and -242; thorium-232; tritium; and uranium-233, -234, -235, and -238 were discharged? What about communities exposed to radioactive waste from the Mallinckrodt Chemical Works, which processed uranium for the US nuclear program and stored waste aboveground north of St. Louis, Missouri? What about other nuclear weapons production sites across the US where fallout was released off-site and people were exposed?

Manhattan Project and Cold War–era production and testing of nuclear weapons by the United States damaged the health and lives of countless civilians. For communities injured by a "nuclear incident" in which the P-AA applies to any effort to pursue a legal remedy, the solution may be to amend the P-AA, setting a cap on defense spending in personal injury litigation in order to "even the playing field." A cap on defense spending would avoid a repetition of the scorched-earth defense waged against the downwinders in the twenty-four-year-long In re Hanford, with $80 million in taxpayer funds expended to defeat downwinders' personal injury claims, in the process trampling the goal of the P-AA to provide efficient compensation for civilians injured by a nuclear incident. And for communities downwind of atomic testing sites for which the Warner Amendment restricts legal options, the solution may be expansion of RECA to compensate individuals who meet radiation exposure requirements and have been diagnosed with diseases recognized as radiogenic, even if their place of residence was geographically distant from the test site or was downwind of waste storage facilities from which radiation was released. Based on failed efforts thus far, however, it is questionable whether Congress will ultimately support the expansion of RECA.

During the Manhattan Project and Cold War era, nuclear weapons production and testing by the US government exposed many American and Marshallese civilians to harmful levels of radioactive fallout. These civilians were lied to and mistreated by the US government. It is time to make things right. This will require that Congress overcome its long-held reticence to provide relief to those deserving relief at the risk of establishing precedent for other similarly injured groups of US civilians. It is time to provide civilian downwinders with compassionate care and compensation for the cancers and other illnesses that, when diagnosed in nuclear workers, are recognized as radiation related. It is time to treat civilians equally and civilly, recognizing their sacrifices and suffering as we have recognized the sacrifices and suffering of America's nuclear workers.

February 18, 1999

Dr. Richard Jackson
Director
National Center for Environmental Health
Centers for Disease Control and Prevention
4770 Buford Highway, NE
Mailstop F29
Atlanta, Georgia 30341-3724

Dear Dr. Jackson:

We are writing to express our profound dismay and objections to the manner and process by which the results of the Hanford Thyroid Disease Study were released last month. The way in which the report was released showed a contemptible lack of sensitivity to the individuals whose personal well-being and family and community health have been, and continue to be, jeopardized by past exposures to Hanford radiation. Moreover, it is already clear that the substantive basis for the report's conclusions is dubious; that uncertainties about the accuracy of the doses assigned to study subjects should have reconciled before such definitive conclusions were offered to the Congress, the press, and the public at large.

We would like to emphasize at the outset that we are not objecting to the news, per se, that an epidemiologic investigation could detect no correlation between exposures and health outcomes. Obviously, such findings are going to occur, more often than not, as scientists test environmental epidemiologic hypotheses with limited observational tools. That's not the issue here.

Our grievance with the Hanford Thyroid Disease Study is that the conclusiveness of the study's findings are not yet warranted by the quality of the science. Officials and scientists at the Centers for Disease Control and Prevention had advance knowledge of these shortcomings and limitations. It is inexplicable that they failed to publicly disclose them. Furthermore, it is inexcusable that they did not seek to explain how the conclusions drawn in the draft report are, at best, premature.

BACKGROUND OF THE HANFORD THYROID DISEASE STUDY

Before addressing the technical shortcomings of the HTDS report, it is important that we address, and you recognize, the harm inflicted on people who were most concerned about the results of this study.

This epidemiologic study was originally requested by the Hanford Education Action League (HEAL) of Spokane, Washington. HEAL came into existence in 1984 after questions arose about whether the U.S. Department of Energy had provided thorough and honest answers to questions about the nature and magnitude of radiation releases from the Hanford plant. Historic documents requested by journalists and public researchers were finally made available in April of 1986, and from these documents the public first learned that large emissions of radioactive iodine had occurred at Hanford during the 1940s and 1950s. When initial dose estimates were provided in the fall of 1986, the magnitude of these initial estimates confirmed the earliest indications that the exposures caused by Hanford emissions were of clear public health concern. Numerous people who had been living in the Hanford environs

Dr. Jackson, page 2

during the period of the radioiodine emissions came forward to publicly identify themselves and family members as having thyroid and other diseases that might be attributable to the radiation exposures. It was on the basis of these public concerns that HEAL petitioned Congress for the funding necessary initiate the HTDS.

The genesis of this study is important because it underscores who the study was for and how the results should have been handled. Certainly, there was expected scientific value in this study in the sense that it was constructed to shed light on the relationship between radioiodine exposures and thyroid disease. But the primary impetus and value of this study was public accountability, as an effort to better quantify the health injuries expected as a result of the Hanford exposures. The last thing that should have happened by the release of this study was to do it in a way that added to the physical and psychological insults already inflicted on Hanford downwinders. But that is precisely the effect of the process by which the study was released.

CDC's FAILURE TO PROVIDE ADEQUATE CONTEXT AND PERSPECTIVE WITH REGARD TO THE STUDY'S RESULTS WAS IRRESPONSIBLE

There are certain things about the Hanford experience that are not at issue. The facts are clear that the exposures occurred. Although there is clear social and scientific value in determining the dynamics of how such exposures affect populations, the question in this instance was never whether the affected population was, and is, at greater health risk on account of being exposed. They were and are. The only question is whether those risks are detectable and quantifiable by methods currently available to epidemiologists.

Scientists and officials at the Centers for Disease Control and Prevention are well aware of the inherent limitations of environmental epidemiologic studies to detect cause and effect relationships between exposures and disease outcomes. They are also well aware, by now, of the social harm that can be done to communities when epidemiologic studies of this kind cannot detect the exposure/harm relationship. The harm is that inclusive or "negative" findings are widely interpreted as discrediting the legitimate concerns of exposed people and the experience of individuals who may very well be correct in their beliefs that their injuries, or the premature deaths of family members and friends, are indeed the result of being exposed. In these instances, people are first injured by polluters who deprive them of their health and peace of mind; and then by scientists who, by failing to adequately communicate the context and limitations of their findings, inflict another helping of emotional distress on such people--in this instance, the thousands of people who identify themselves as Hanford downwinders.

Given the high national and regional profile of this study, it is deplorable how little was done to prepare the media, the public, and the exposed population for the release of this report. Ironically, early copies of the report were marked "SENSITIVE." But to whom was the sensitivity owed? CDC? The Department of Energy? The Congress? No. The sensitivity was owed to the people whose health and whose emotional investment in their experience as exposed persons was at stake. And yet, the way in which the report was released and presented displayed a callous indifference to the well-being of these individuals.

Dr. Jackson, page 3

CDC OFFICIALS SHOULD HAVE INTERVENED TO CORRECT THE CONTRACT SCIENTISTS WHEN THE SIGNIFICANCE OF THE HTDS RESULTS WAS CLEARLY OVERSTATED

Even before this report was publicly presented in Richland on the evening of January 28th, CDC officials and scientists had to have known that the contract investigators were publicly overstating both the significance of their findings and the confidence these findings deserve. Yet, not only did they arrange for the release of this report in a manner that virtually guaranteed that exposed individuals would first read about it in a national newspaper, but they stood by silently and loaned their credibility to assertions by the contract scientists that are simply irresponsible.

Given the unresolved technical issues discussed below, we take strong issue with the statement from the report's executive summary that,

"[T]hese results provide rather strong evidence that exposures at these levels to I-131 do not increase the risk of thyroid disease or hyperparathyroidism. These results should consequently provide a substantial degree of reassurance to the population exposed to Hanford radiation that the exposures are not likely to have affected their thyroid or parathyroid health."

Frankly, the strength of the evidence remains to be demonstrated. More importantly, if we put the preliminary results of the HTDS in the context of previous epidemiologic investigations looking at radiation effects to the thyroid (including I-131) it would be much more reasonable to conclude that Hanford radiation did, in fact, affect the health of some number of individuals in the exposed population. The appropriate message warranted by the preliminary results of the HTDS is that the researchers were unable to positively correlate the exposures to the incidence of thyroid disease in that portion of the exposed population included in the HTDS study. That should have been the message. No more. No less.

KEY TECHNICAL CONCERNS AFFECTING THE RELIABILITY OF THE HTDS RESULTS SHOULD HAVE BEEN DISCLOSED AND ADDRESSED

There were at least two technical concerns about the HTDS that should have been addressed when the report was released and were not.

The first is the issue of statistical power. As you know, the power of a study directly determines the anticipated degree of confidence with which investigators can accept the null hypothesis--in this case the hypothesis that there was and is no correlation between thyroid doses attributable to Hanford iodine-131 and the incidence of thyroid disease in the HTDS cohort population. Since the inception of the HTDS, the issue of adequate power has been a concern as it is in any such study where the number of study subjects is limited and the knowledge of the doses received is skewed heavily toward the low end, where any expected correlation between exposure and disease outcome is expected to be weak.

As the National Research Council's Committee on the Assessment of CDC's Radiation Studies concluded in a November 16, 1994 letter to CDC about a review of progress and plans for the HTDS: *"The committee's main reservations were that the statistical power, although adequate, is not outstanding, and that some questions about dosimetry remain unresolved."*

Dr. Jackson, page 4

When the results of the HTDS were made public on January 28th, the HTDS investigators reported, in the words of principal investigator Scott Davis, that "it was important to point out, in considering these [results], the study is a powerful study."

There was no discussion, by either the investigators or the CDC officials and scientists present during the briefings, about how the critical statistical power issues had been resolved, and whether CDC concurred that they had been resolved. Notably absent was any discussion about how uncertainties in the dose estimates could affect the confidence of the regression analysis (the critical dose response function).

The most egregious omission, however, was the total lack of disclosure and candor having to do with the critical uncertainties in the doses assigned to individuals in the HTDS cohort.

In a letter transmitted to CDC just 10 days before the HTDS was released, the NRC Committee on the Assessment of CDC's Radiation Studies, raised and emphasized problems with the uncertainties of individual doses calculated with the Hanford Environmental Dose Reconstruction methods used in conjunction with the HTDS study.

"[I]t should be noted," the Committee reported, "that the inherent uncertainty associated with the individual doses will decrease the likelihood of determining a meaningful risk coefficient for the effects of radioiodine on the target population."

In the Sunday, February 14 edition of the Spokane, Washington Spokesman-Review, Scott Davis, the principal HTDS investigator, is reported to say that "he couldn't agree more" that there should have been a more thorough scientific review of the study before it was released.

It is appalling that CDC would go forward with the release of the HTDS under such circumstances, and so quickly after its NRC review committee had identified such major problems.

Unfortunately, it will be extremely difficult to repair the harm done to CDC's credibility as a result of this fiasco. It is simply hard to imagine that any community in America, concerned about environmental exposures to pollutants and subsequent health outcomes, would welcome CDC to come in and conduct, sponsor, or otherwise oversee and epidemiologic study. We are deeply troubled by this.

In the short run, CDC should make plans to set the record straight before Congress, the news media, the people of the Pacific Northwest, and especially the Hanford downwinders. We also suggest that a public apology is in order.

The sooner these actions are taken the better. It is also our expectation that CDC will provide the Advisory Committee on Energy-Related Epidemiologic Research, at its April 1999 meeting, a progress report on the steps taken to correct the record both with regard to the limitations of the HTDS study and the unwarranted conclusions that were presented to Congress, the press and the public on January 27th and 28th.

Sincerely,

Susan Guidroz in SSN'
NIESH, Inc.

Edna Yocum
RESH, Inc.

Dr. Tillie B. Poteet
Stand
Dr. Fay M. Martin

Tim Connor

Trisha E. Pritikin

Tom J. Baker (CFORVO)

Patrice Sutton
Dr. Seth Tuler (CCRI)

Dr. Jackson, page 5

Additional Signatures

[signature] Idaho J. Lindsa

[signature] Geoffrey Sea

[signature] Rachel Moses

[signature] Richard Bandy (TEWA)

[signature] Linda Kerr (HHES)

[signature] Marcia Wood (HHES)

[signature] Martha Hoskling (Schwick Tribe)

[signature] Judith Jurji (HDC, HHES)

[signature] Lynne M Stembridge (HEAL)

[signature] Robert Walls

[signature] Madeline Cunningham Spino (Warm Springs Tribe)

[signature] DJ Jim; Yakama Nation Environmental Prog
P.O. Box 151, Toppenish, WA 98948

[signature] Linda Kerr (Northwest Radiation Health Alliance)

[signature] Louie Yc (HHES)

Additional Electronic Signatures

Tom Marshall, Rocky Mountain Peace & Justice Center, Boulder, Colorado
Preston J. Truman, Downwinders, Lava Hot Springs, Idaho
Chuck Broscious, Environmental Defense Institute, Troy Idaho
Alice Slater, Global Resource Action Center for the Environment, New York, NY
Jana Gouge, Harriman, Tennessee

NOTES

1. THE FORGOTTEN GUINEA PIGS

1. The Atomic Energy Act of 1954 authorized the clandestine production of atomic weapons by government-controlled industry free of public oversight and exempt from most local, state, and federal environmental and health regulations. See Michael D'Antonio, *Atomic Harvest: Hanford and the Lethal Toll of America's Nuclear Arsenal* (New York: Crown, 1993), 207.

2. There are two types of nuclear bomb. All atomic bombs are nuclear bombs, but not all nuclear bombs are atomic bombs. Atomic bombs depend on fission, with their explosive energy derived from splitting atoms in materials such as uranium or plutonium. Hydrogen bombs (also known as thermonuclear bombs) depend upon the fusion of atoms, releasing far greater energy than atomic bombs. Atomic bombs are generally used as triggers for thermonuclear bombs. See Brajendra Singh, "What Is the Difference between an Atomic and a Nuclear Bomb?," *Times of India*, December 24, 2005.

3. Stephanie Buck, "Fear of Nuclear Annihilation Scarred Children Growing Up in the Cold War Era, Studies Later Showed," *Timeline*, August 29, 2017.

4. Author interview with Bonnie Rae, June 11, 2018. Rae's mother worked at Lady of Lourdes Hospital in the 1950s and 1960s and frequently mentioned the high number of children with leukemia coming to the hospital for treatment.

5. US Congress, House Committee on Interstate and Foreign Commerce, Subcommittee on Oversight and Investigations, *The Forgotten Guinea Pigs: A Report on Health Effects of Low-Level Radiation Sustained as a Result of the Nuclear Weapons Testing Program Conducted by the United States Government*, 96th Cong., 2nd sess., August 1980.

6. Workers in the uranium industry also face the risks of lung cancer, leukemia, birth defects in their offspring, and other serious health issues. See Nathan Rice, "Health Legacy of Uranium Mining Lingers Thirty Years Later," *Scientific American*, June 28, 2010.

7. From the title of Harvey Wasserman and Norman Solomon with Robert Alvarez and Eleanor Walters, *Killing Our Own: The Disaster of America's Experience with Atomic Radiation* (New York: Delacorte, 1982).

2. HANFORD AND THE MANHATTAN PROJECT

1. "B Reactor Background and History," Hanford Site, https://www.hanford.gov/page.cfm/BReactor.

2. Michele Stenehjem Gerber, *On the Home Front: The Cold War Legacy of the Hanford Nuclear Site* (Lincoln: University of Nebraska Press, 1992), 33.

3. S. L. Sanger, *Working on the Bomb* (Portland, OR: Portland State University Press, 1995), 36n.

4. Sanger, 149.

5. "B Reactor Background and History."

6. Gerber, *On the Home Front*, 34.

7. Daniel Grossman, "Hanford and Its Early Atmospheric Releases," in "The Nuclear Northwest," *Pacific Quarterly* 85, no. 1 (January 1994): 6–14, 8.

8. Grossman, 10.

9. Sanger, *Working on the Bomb*, 148.

10. Leslie M. Groves, *Now It Can Be Told: The Story of the Manhattan Project* (1962; repr., New York: Da Capo Press, 1983), 69. Groves writes about the graphite reactor, a plutonium production reactor, at the Manhattan Project's Clinton Site, Oak Ridge, Tennessee, and the possible dangers to nearby Knoxville, a major population center, from an explosion of the reactor. The Clinton graphite reactor was the pilot reactor that demonstrated the feasibility of large-scale plutonium production in the reactors at Hanford. These early safety concerns led to the selection of the Hanford site as the primary plutonium production facility for the Manhattan Project. DuPont, the initial contractor to operate Hanford, reported, "The early experience at Clinton Laboratories sufficed to define rather clearly the radiation hazards to be expected in the separations process at Hanford." HAN-73214, E. I. du Pont de Nemours and Co., Book VII, 87.

11. Gerber, *On the Home Front*, 102.

12. Gerber, 102.

13. Michele Stenehjem Gerber, "A Brief History of the T-Plant Facility at the Hanford Site," May 1994, WHC-MR-0452, Addendum 1, University of Washington Libraries, Special Collections, Hanford Litigation Records.

14. The individuals whose stories appear in this book were plaintiffs in *In re Hanford Nuclear Reservation Litigation*. Stories are presented chronologically based on when individuals were exposed to Hanford radiation. Some stories have been edited for clarity.

15. "Juvenile Rheumatoid Arthritis (JRA)," http://www.houstonrheumatology.com/pdfs/conditions/juvenile-rheumatoid-arthritis.pdf.

16. "Juvenile Rheumatoid Arthritis (JRA)."

17. See Rasoul Yahyapour, Peyman Amini, Saeed Rezapour, Mohsen Cheki, Abolhasan Rezaeyan, Bagher Farhood, Dheyauldeen Shabeeb, Ahmed Eleojo Musa, Hengameh Fallah, and Masoud Najafi, "Radiation-Induced Inflammation and Autoimmune Diseases," *Military Medical Research* 5, no. 2 (March 2018): 9.

18. O. Axelson, A. M. Landtblom, and U. Flodin, "Multiple Sclerosis and Ionizing Radiation," *Neuroepidemiology* 20, no. 3 (August 2001): 175–178.

19. Walter B. Eidbo and Merle P. Prater, "Linkage—Multiple Sclerosis and Ionizing Radiation," http://www.wpb-radon.com/Radon_research_papers/2004%20 Newport,%20RI/2004_04_Linkage%20Multiple%20Sclerosis%20and%20Ioniz ing%20Radiation.pdf.

20. Axelson, "Multiple Sclerosis and Ionizing Radiation."

21. Dahr Jamail, "Fukushima Radiation Alarms Doctors," *Al Jazeera*, August 18, 2011.

22. Eiichiro Ochiai, "The Manga 'Oishinbo' Controversy: Radiation and Nose Bleeding in the Wake of 3.11," *Asia-Pacific Journal*, 11, issue 25, no. 4 (June 23, 2013): 1–11.

23. Ochiai.

24. Low doses are defined as those ranging from near zero to about 100 mSv of low-LET (linear energy transfer) ionizing radiation. National Academies of Science, *Biologic Effects of Ionizing Radiation (BEIR) VII, Phase 2: Health Risks from Exposure to Low Levels of Ionizing Radiation* (Washington, DC: National Academies Press, 2006), 2.

25. R. Wakeford, "The Meaning of Low Dose and Low Dose-Rate," *Journal of Radiological Protection* 30, no. 1 (March 2010): 1–3.

26. D. B. Shipler et al., "Hanford Environmental Dose Reconstruction Project," *Health Physics* 71, no. 4 (October 1996): 532–544.

27. "Analytical Report of Radioactive Substances in Special Urine Samples," September 24, 1954, HW 33145. "HW" refers to "Hanford Works" or "Hanford Engineer Works," early terms for the Hanford site. Many HW documents are found at https://www.osti.gov/opennet/detail?osti-id=16436939 or within the Declassified Document Retrieval System catalogue of the DOE Public Reading Room in Richland, WA.

28. "Achievements in HAPO Radiation Monitoring, 1944 to 1954," September 15, 1954, HW 33533.

29. "Urine Samples," April 16, 1946, HW 73886, 1.

30. "Radiation Protection Operation: The Routine Detection of Plutonium Deposition Based on Single Void Urine Samples," January 18, 1960, HW 63561.

31. "Urine Samples," 3.

32. "Urine Samples," 2.

33. Murray Campbell, "Project Sunshine's Dark Secret," *Globe and Mail* (Canada), April 12, 2018.

34. "Project Sunshine," *Chicago Bulletin* no. 11, December 21, 1955, box 2242, US DOE Archive, National Archives and Records Administration, College Park, MD.

35. Campbell, "Project Sunshine's Dark Secret."

36. ACHRE, "Documentary Update on Project Sunshine 'Body Snatching,'" June 9, 1995, ACHI.000015.004b, https://bioethicsarchive.georgetown.edu/achre/com meet/meet15/brief15/tab_d/br15d2.html.

37. Campbell, "Project Sunshine's Dark Secret."

38. Ker Than, "First Atomic Bomb Test Exposed US Civilians to Radiation," Live Science, July 16, 2007, https://www.livescience.com/1698-atomic-bomb-test -exposed-civilians-radiation.html.

39. Memorandum to Major General Groves from Colonel Stafford L. Warren, Chief of Medical Section, Manhattan District, National Archives, Record Group 77, Records of the Office of the Chief of Engineers, Manhattan Engineer District, TS Manhattan Project Files, folder 4, "Trinity Test."

40. Press release, Office of the Commanding Officer, Alamogordo Base, July 16, 1945, in Groves, Now It Can Be Told, 301.

41. September 2, 1945, the day the Japanese delegation formally signed the instrument of surrender onboard the USS Missouri, is recognized as the official end of World War II.

42. Gerber, On the Home Front, 2.

43. Gerber, On the Home Front, 65.

44. H.E.W., the abbreviation for Hanford Engineer Works, was the name given to the Hanford project by the US Army Corps of Engineers and its initial contractor, DuPont. It was renamed Hanford Works (HW) when the AEC assumed control of the site on January 1, 1947. General Electric, the contractor that succeeded DuPont, called the site the Hanford Atomic Products Operation (HAPO).

45. Herbert M. Parker and Simon T. Cantril, "The Status of Health and Protection at Hanford Engineer Works," August 24, 1945, HW 72136, 6, University of Washington Libraries, Special Collections, Hanford Litigation Records.

46. Parker and Cantril, 6.

47. Gerber, On the Home Front, 2.

48. "I-131 Accumulation in the Thyroid of Sheep Grazing Near H.E.W.," March 1, 1946, HW 33455, 1.

49. "I-131 Deposition in Cattle Grazing on North Margin of H.E.W.," August 29, HW 33628, 1.

50. "I-131 Accumulation in the Thyroid of Sheep Grazing Near H.E.W.," March 1, 1946, HW 33455, 2.

51. Kate Brown, Plutopia: Nuclear Families, Atomic Cities, and the Great Soviet and American Plutonium Disasters (Oxford, UK: Oxford University Press, 2013), 166.

52. "I-131 Deposition in Cattle Grazing on North Margin of H.E.W.," August 29, 1946, HW 33625, 1.

53. "Evaluation of Radiological Conditions in the Vicinity of Hanford," January–June 1964, August 17, 1964, HW 83723, 5.

54. Gerber, *On the Home Front*, 109.

55. Brown, *Plutopia*, 226.

3. THE EARLY COLD WAR, 1945–1950

1. DOE, "Hanford, Cold War Forever Linked: Site's Defense Mission Continued Long after the End of World War II," *Tri-City Herald* (Kennewick, WA), October 28, 2009.

2. The PPG comprised the Marshall Islands District of the Trust Territory of the Pacific Islands, administered and controlled by the United States. The United States conducted 105 atmospheric and underwater tests from 1946 to 1962 within the PPG. See nuclearweaponarchive.org.

3. Kat Eschner, "The Crazy Story of the 1946 Bikini Atoll Nuclear Tests," *Smithsonian* magazine, June 30, 2017, https://www.smithsonianmag.com/smart-news /crazy-story-1946-bikini-atoll-nuclear-tests-180963833.

4. Atomic Energy Act of 1946 (McMahon Act), ch. 724, § 2(a)(1). 60 Stat. 755, 756, repealed by Energy Reorganization Act of 1974, Pub. L. No. 93-438, § 104(a), 88 Stat. 1233, 1237.

5. "Civilian Control of Atomic Energy," The Manhattan Project, https://www .osti.gov/opennet/manhattan-project-history/Events/1945-present/civilian _control.htm. The Manhattan District was abolished on August 15, 1947.

6. See "How Do Nuclear Weapons Work?," Union of Concerned Scientists, https://www.ucsusa.org/nuclear-weapons/how-do-nuclear-weapons-work# .W39PDn4na7o.

7. "Berlin Blockade," History, https://www.history.com/topics/cold-war/berlin -blockade.

8. "The Chinese Revolution of 1949," Department of the Historian, US Department of State, https://history.state.gov/milestones/1945-1952/chinese-rev.

9. In 1958, Richland was incorporated as a chartered, self-governed first-class city. Government ownership ended, and houses were then available for purchase.

10. "Hyperparathyroidism," Mayo Clinic, http://www.mayoclinic.org/diseases -conditions/hyperparathyroidism/basics/definition/con-20022086.

11. N. Takeichi, K. Dohi, H. Ito, H. Yamamoto, K. Mabuchi, T. Yamamoto, K. Shimaoka, and K. Yokoro, "Parathyroid Tumors in Atomic Bomb Survivors in Hiroshima: Epidemiological Study from Registered Cases at Hiroshima Prefecture

Tumor Tissue Registry, 1974–1987," *Japanese Journal of Cancer Research* 82 (August 1991): 875–878.

12. A. B. Schneider, T. C. Gierlowski, E. Shore-Freedman, M. Stovall, E. Ron, and J. Lubin, "Dose-Response Relationships for Radiation-Induced Hyperparathyroidism," *Journal of Clinical Endocrinology and Metabolism* 80, no. 1 (January 1995): 254–257.

13. Notice of Regulations: Compensation for Damage to Person, Attachment 1, RCC1.951299.002A. See also *Marshall Islands Nuclear Claims Tribunal Act 1987*, 42 MIRC Ch. 1, http://rmiparliament.org/cms/images/LEGISLATION/PRINCIPAL /1987/1987-0024/MarshallIslandsNuclearClaimsTribunalAct1987_1.pdf.

14. C. A. Fink and M. N. Bates, "Melanoma and Ionizing Radiation: Is There a Causal Relationship?," *Radiation Research* 164, no. 5 (November 2005): 701–710; R. Wakeford, "The Cancer Epidemiology of Radiation," *Oncogene*, 23, no. 38 (August 2004): 6404–6428; D. H. Moore, H. W. Patterson, F. Hatch, D. Discher, J. S. Schneider, D. Bennett, and M. L. Mendelsohn, "Case-Control Study of Malignant Melanoma among Employees of the Lawrence Livermore National Laboratory," *American Journal of Industrial Medicine* 32, no. 4 (October, 1997): 377–391; D. M. Freeman, A. Sigurdson, R. S. Rao, M. Hauptmann, B. Alexander, A. Mohan, M. Morin Doody, and M. S. Linet, "Risk of Melanoma among Radiologic Technologists in the United States," *International Journal of Cancer* 103 (February 2003): 556–562.

15. See Keith Schneider, "Washington Nuclear Plant Poses Risk for Indians," *New York Times*, September 3, 1990.

16. "The Kalispel Tribal Health Survey—2006," http://www2.clarku.edu/mta fund/prodlib/kalispel/06-Survey_Final_Report.pdf.

17. "Hanford Exposed Tribes, Study Says," *Herald Net* (Everett, WA), January 24, 2002, https://www.heraldnet.com/news/hanford-exposed-tribes-study-says.

18. Associated Press, "Law Firms Sue Hanford on Behalf of Indians: Suit against Government Contractors, UW Claims Indians Deliberately Exposed to Radiation," *Spokesman-Review* (Spokane, WA), April 6, 1997.

19. Lisa Stiffler, "Toxic Fish Imperil Tribes," *Seattle Post-Intelligencer*, July 31, 2002.

20. Stiffler.

21. Dave Birkland, "More Hanford Radiation? Indians May Have Higher Exposure Than Previously Estimated," *Seattle Times*, January 25, 2002.

22. Robert Alvarez, "The Legacy of Hanford: Washington Continues to Evade Responsibility for Forty-Seven Years of Contamination," *Nation*, July 31, 2003.

23. William Willard and Mary Kay Duffe, *Final Report: Migrant Farm Worker Scoping Study*, March 1, 1997 (available on request from US Agency for Toxic Substances and Disease Registry).

24. Michele Stenehjem Gerber, *On the Home Front: The Cold War Legacy of the Hanford Nuclear Site* (Lincoln: University of Nebraska Press, 1992), 104.

25. Jim Kershner, "Construction of North Richland Construction Camp, Which Will Eventually House 25,000 Hanford Workers and Their Families, Is Authorized on August 15, 1947," HistoryLink.org, June 25, 2014, https://www.histo rylink.org/File/10805.

26. "Secrets, Lies and Operation Bluenose," Environmental Defense Institute, October 2000, http://www.environmental-defense-institute.org/publications /NEWS.Oct.00.html.

27. See Karen Dorn Steele, "Hanford's Bitter Legacy," *Bulletin of Atomic Scientists* (January/February 1988): 20.

28. Hanford's T plant was the first chemical processing and separation plant of its kind in the world.

29. "Human Radiation Studies: Remembering the Early Years. Oral History of John W. Healy," interview by the US DOE Office of Human Radiation Experiments, November 2, 1994, DOE/EH-0455.

30. J. W. Healy et al., "Dissolving of Twenty Day Metal at Hanford," May 1, 1950, HW 17381.

31. Michele Stenehjem Gerber, "A Brief History of the T-Plant Facility at the Hanford Site," 32, WHC-MR-0452, University of Washington Libraries, Special Collections, Hanford Litigation Records.

32. "Human Radiation Studies: Remembering the Early Years. Oral History of Health Physicist Carl C. Gamertsfelder," interview by US DOE Office of Human Radiation Experiments, January 19, 1995.

33. Michael D'Antonio, *Atomic Harvest: Hanford and the Lethal Toll of America's Nuclear Arsenal* (New York: Crown Publishing, 1993), 125.

34. Karen Dorn Steele, "'49 Iodine Release Went Awry: Radioactive Gases Were Two to Three Times More Than Intended," *Spokane Chronicle*, May 3, 1989, 1.

35. US General Accounting Office, "Nuclear Health and Safety: Examples of Post–World War II Radiation Releases at US Nuclear Sites," Washington, DC, November 1993, 10.

36. Weather results for Walla Walla, WA, February 3, 1949, Farmers' Almanac, http://farmersalmanac.com/weather-history/99362/1949/02/03.

37. Hanford Health Information Network, *Hot Spots: Weather and Hanford's Radiation Releases to the Air*, September 1994, Washington State Digital Archives, Cheney, WA.

38. (Colonel and Radiological Safety Officer) James P. Cooney, Memorandum to Rear Admiral Parsons, May 11, 1948, Annex "B" in US Commanding Lieutenant

General J. E. Hull's memorandum to US Army Chief of Staff, "Subject: Location of Proving Ground for Atomic Weapons." See "Radioactive Heaven and Earth," Institute for Energy and Environmental Research, 1991, https://ieer.org/wp/wp-content/uploads/1991/06/RadioactiveHeavenEarth1991.pdf.

39. Gordon Eliot White, "Nevada A-Site Picked in a Hurry," *Deseret News* (Salt Lake City, UT), December 8, 1978.

40. Atomic Energy Commission, *Report 141/7, Meeting No. 504* (December 2, 1950). National Archives and Records Administration, College Park, MD.

41. People living downwind of Hanford may have been exposed as well to fallout from testing at the NTS and from global tests that drifted over the Hanford area. An online dose calculator, sponsored by the US National Cancer Institute, allows Hanford downwinders and others to calculate their additional I-131 exposures from these sources along with the resultant risks of developing thyroid cancer from those exposures: https://radiationcalculators.cancer.gov/fallout.

4. THE COLD WAR, 1951–

1. *Allen v. United States*, 588 F. Supp. 247, 372 (D. Utah 1984).

2. Philip L. Fradkin, *Fallout: An American Nuclear Tragedy* (1989; repr., Boulder, CO: Johnson Books, 2004), 183–184.

3. Fradkin, 183.

4. Harold L. Beck, Lynn R. Anspaugh, André Bouville, and Steven L. Simon, "Review of Methods of Dose Estimation for Epidemiological Studies of the Radiological Impact of Nevada Test Site and Global Fallout," *Radiation Research* 166, no. 1, part 2 (July 2006): 209–218. See also *Allen v. United States*, 588 F. Supp. 247, 372.

5. *Allen v. United States*, 588 F. Supp. 247, 372.

6. Gordon Eliot White, "US Kept Ignoring Evidence about Fallout's Deadly Effects," *Deseret News*, October 28, 1990.

7. White.

8. *Allen v. United States*, 588 F. Supp. 247, 378.

9. *Allen v. United States*, 588 F. Supp. 247, 380.

10. Nick Knight, "A Trawler Was Caught in the Fallout of a Nuclear Test, Killing Its Radioman; His Last Words Were "I Pray That I Am the Last Victim of an Atomic or Hydrogen Bomb," Vintage News, August 11, 2016, https://www.thevintagenews.com/2016/08/11/japanese-trawler-caught-fallout-castle-bravo-nuclear-test-killing-radioman-last-words-pray-last-victim-atomic-hydrogen-bomb.

11. Knight.

12. Matthew L. Wald, "Kodak Warned of Fallout Threat: Government Regularly

Alerted Film Manufacturers While Telling Public There Was No Danger," *Spokesman-Review* (Spokane, WA), September 30, 1997.

13. Ernest Sternglass, *Secret Fallout: Low-Level Radiation from Hiroshima to Three Mile Island* (New York: McGraw-Hill, 1981), 13.

14. Sternglass, 5.

15. *Annual Report of the Health Instrument Divisions 1950*, July 20, 1951, HW 21699, 6.

16. *Quarterly Progress Report Research and Development Activities for April–June 1950*, July 19, 1950, HW 18371, 13.

17. *Annual Report of the Health Instrument Divisions 1950*, 41.

18. The abbreviation μCi/d represents microcuries per day. A microcurie is a quantity of radioactivity equivalent to one-millionth of a curie.

19. L. K. Bustad, L. A. George, S. Marks, D. E. Warner, C. M. Barnes, K. E. Herde, and H. A. Kornberg, "Biological Effects of I-131 Continuously Administered to Sheep," *Radiation Research* 6, no. 3 (March 1957): 380–413, 382.

20. Bustad et al., 382

21. Bustad et al., 396.

22. Bustad et al., 396.

23. Bustad et al., 395–396.

24. Bustad et al., 398.

25. Bustad et al., 398.

26. Bustad et al., 397.

27. Bustad et al., 397.

28. *Quarterly Progress Report Research and Development Activities for April–June 1950*, 14.

29. Bustad et al., "Biological Effects of I-131," 396.

30. *Quarterly Progress Report, July–September, 1950*, October 16, 1950, HW 19146, 12.

31. Bustad et al., "Biological Effects of I-131," 396–397.

32. Bustad et al., 397.

33. Bustad et al., 401.

34. Bustad et al., 397.

35. Bustad et al., 397.

36. Radiological Sciences Department, *Quarterly Progress Report Jan.–March 1952*, April 16, 1952, HW 24131, 12.

37. Bustad et al., "Biological Effects of I-131," 406.

38. Bustad et al., 397.

39. Bustad et al., 397.

40. *Biology Research Annual Report 1956*, January 4, 1957, HW 47500, 9.

41. *Biology Research Annual Report 1956*, 96.

42. Michele Stenehjem Gerber, *On the Home Front: The Cold War Legacy of the Hanford Nuclear Site* (Lincoln: University of Nebraska Press, 1992), 97.

43. Gerber, 97.

44. US Congress, House Committee on Interstate and Foreign Commerce, Subcommittee on Oversight and Investigations, *Low-Level Radiation Effects on Health*, 96th Cong., 1st sess., April 23, May 24, August 1, 1979, 525 (hereafter cited as *Low-Level Radiation Effects*).

45. *Low-Level Radiation Effects*, 526.

46. US Congress, Joint Hearing of the House Committee on Interstate and Foreign Commerce, Subcommittee on Oversight and Investigations, and the Labor and Human Resources Committee, Health and Scientific Subcommittee, and the Committee on the Judiciary, *Health Effects of Low-Level Radiation*, 96th Cong., 1st sess., April 19, 1979, vol. 2, 1431 (hereafter *Health Effects of Low-Level Radiation*).

47. *Health Effects of Low-Level Radiation*, vol. 2, 1431.

48. US Congress, House Committee on Interstate and Foreign Commerce, Subcommittee on Oversight and Investigations, *The Forgotten Guinea Pigs: A Report on Health Effects of Low Level Radiation Sustained as a Result of the Nuclear Weapons Testing Program Conducted by the United States Government*, 96th Cong., 2nd sess., August 1980, 5 (hereafter cited as *Forgotten Guinea Pigs*).

49. *Health Effects of Low-Level Radiation*, vol. 2, 1431.

50. *Health Effects of Low-Level Radiation*, vol. 2, 1431.

51. *Health Effects of Low-Level Radiation*, vol. 2, 1431.

52. *Health Effects of Low-Level Radiation*, vol. 2, 1442.

53. *Health Effects of Low-Level Radiation*, vol. 2, 1432.

54. *Health Effects of Low-Level Radiation*, vol. 2, 1429.

55. *Health Effects of Low-Level Radiation*, vol. 2, 1430.

56. *Health Effects of Low-Level Radiation*, vol. 2, 1435.

57. Jim Kichas, "Downwind in Utah," Utah Division of Archives and Records Service, June 26, 2015, https://archivesnews.utah.gov/2015/06/26/downwind-in-utah.

58. *Forgotten Guinea Pigs*, 5.

59. *Health Effects of Low-Level Radiation*, vol. 1, 98.

60. Kichas, "Downwind in Utah."

61. *Bulloch v. United States*, 145 F. Supp. 824 (D. Utah 1956), 826.

62. Sarah Alisabeth Fox, *Downwind: A People's History of the Nuclear West* (Lincoln: University of Nebraska Press, 2014), 55.

63. The first suit filed in 1954 by Utah sheep ranchers for loss of their herds,

Bulloch v. United States, 145 F. Supp. 824, is referred to as Bulloch I. The second suit filed by the same sheep farmers decades later and heard by the same judge, Bulloch v. United States, 95 Fed. Rules Dec. 123 (1982), is referred to as Bulloch II. In Bulloch II, following four days of testimony, Judge Christensen determined that a species of fraud had been committed by the government on the court in Bulloch I, vacated his prior judgment, ordered the government to pay costs, and scheduled a new trial.

64. Federal Tort Claims Act (FTCA), 28 U.S.C. §§ 2671 et seq.

65. FTCA.

66. Bulloch v. United States, 145 F. Supp. 824, 825. The discretionary function exception to the FTCA is found at 28 U.S.C. § 2680(a).

67. Bulloch v. United States, 145 F. Supp. 824, 825.

68. Bustad et al., "A Comparative Study of Hanford and Utah Range Sheep," HW 30119, November 30, 1953, https://www.osti.gov/opennet/detail?osti-id=16435551.

69. Kate Brown, Plutopia: Nuclear Families, Atomic Cities, and the Great Soviet and American Plutonium Disasters (Oxford, UK: Oxford Press, 2015), 230.

70. Bustad et al., "A Comparative Study of Hanford and Utah Range Sheep," 43.

71. Bustad et al., "A Comparative Study of Hanford and Utah Range Sheep," 1.

72. Bustad et al., "A Comparative Study of Hanford and Utah Range Sheep," 43.

73. Bustad et al., "A Comparative Study of Hanford and Utah Range Sheep," 44.

74. Bustad et al., "A Comparative Study of Hanford and Utah Range Sheep," 43.

75. Bulloch v. United States. 145 F. Supp 824, 828.

76. AEC, Atomic Test Effects in the Nevada Test Site Region, January 1955, https://www.fourmilab.ch/etexts/www/atomic_tests_nevada.

77. AEC, Atomic Test Effects in the Nevada Test Site Region.

78. "What Color Was the Mushroom Cloud?" Peace Seeds, http://www.hiroshimapeacemedia.jp/hiroshima-koku/en/exploration/index_20080128.html.

79. Robert A. Jacobs, The Dragon's Tail: Americans Face the Atomic Age (Amherst: University of Massachusetts Press, 2010), 85.

80. Laura Bliss, "Atomic Tests Were a Tourist Draw in 1950s Las Vegas," CityLab, August 8, 2014, https://www.citylab.com/equity/2014/08/atomic-tests-were-a-tourist-draw-in-1950s-las-vegas/375802.

81. Bliss.

82. Howard Ball, "Downwind from the Bomb," New York Times Magazine, February 9, 1986.

83. Ball.

84. Rory Carroll, "Hollywood and the Downwinders Still Grapple with Nuclear Fallout," Guardian, June 6, 2015.

85. See, e.g., Lawrence S Wittner, "Blacklisting Schweitzer," *Bulletin of the Atomic Scientists* 51, no. 3 (May–June 1995): 55–61.

86. See also Joseph Trevithick, "During the 1950s, the Pentagon Played War Games with Troops and Nukes," Medium, April 20, 2015, https://medium.com/war-is-boring/during-the-1950s-the-pentagon-played-war-games-with-troops-and-nukes-9f66a6bca5b8.

87. See Glyn Caldwell, Matthew M. Zack, Michael T. Mumma, Henry Falk, Clark W. Heath, Jr., John E. Till, Heidi Chen, and John D. Boice, Jr., "Mortality among Military Participants at the 1957 PLUMBBOB Nuclear Weapons Test Series and from Leukemia among Participants at the SMOKY Test," *Journal of Radiological Protection* 36, no. 3 (September 2016): 474–489.

88. US Congress, Joint Committee on Atomic Energy, Special Subcommittee on Atomic Energy, *The Nature of Radioactive Fallout and Its Effects on Man,* 85th Cong., 1st sess., May 27, 28, 29, and June 3, 1957 (hereafter cited as *The Nature of Radioactive Fallout*).

89. Fradkin, *Fallout,* 62–63 (referencing *The Nature of Radioactive Fallout*).

90. US Department of State, Treaty Banning Nuclear Weapon Tests in the Atmosphere, in Outer Space, and under Water, October 10, 1964, https://www.state.gov/t/isn/4797.htm.

91. Preparatory Commission for the Comprehensive Nuclear-Test-Ban Treaty Organization (CTBTO), https://www.ctbto.org/specials/testing-times/18-december-1970-the-baneberry-incident.

5. POISONED MILK

1. James P. Thomas, "Hot Milk: The Discovery of the Milk-Iodine Pathway," University of Washington Libraries, Special Collections, James P. Thomas papers.

2. Thomas, 6.

3. Thomas, citing J. F. McClendon, R. E. Remington, H. von Kolnitz, and R. Rufe, "The Determination of Traces of Iodine. III. Iodine in Milk, Butter, Oil and Urine," *Journal of the American Chemical Society* 52 (February 1930): 541–549, 544.

4. Plaintiffs' Revised Statement of Facts, *In re Hanford Nuclear Reservation Litigation,* U.S. District Court, E.D. Wash., Case No. CV-91-3015-WFN (hereafter cited as Plaintiffs' Revised Statement of Facts), 2, 3, 6; Westinghouse Revised Statement of Facts, *In re Hanford Nuclear Reservation Litigation,* 6, University of Washington Libraries, Special Collections, Hanford Litigation Records.

5. Plaintiffs' Revised Statement of Facts, 18, 19, 20, and 22.

6. Plaintiffs' Revised Statement of Facts, 21.

7. Simon Cantril and John W. Healy, "Iodine Metabolism with Reference to I-131," October 22, 1945, HW 72604, University of Washington Libraries, Special Collections, Hanford Litigation Records.

8. Plaintiffs' Revised Statement of Facts, 9–11.

9. Cantril and Healy, "Iodine Metabolism with Reference to I-131," 2–3.

10. Cantril and Healy, 3.

11. Cantril and Healy, 3.

12. *Richland (WA) Villager*, November 25, 1945, 3.

13. Biology Research—Annual Report 1953, HW 30437, January 4, 1954, 138.

14. Defendants' Revised Statement of Facts, *In re Hanford Nuclear Reservation Litigation*, 20, University of Washington Libraries, Special Collections, Hanford Litigation Records.

15. Defendants' Statement of Facts, *In re Hanford Nuclear Reservation Litigation* (hereafter cited as Defendants' Statement of Facts), 12–14, University of Washington Libraries, Special Collections, Hanford Litigation Records.

16. Defendants' Statement of Facts, 15.

17. *Allen v. United States*, citing W. Johnson, et al., "Monitoring of Cow's Milk for Fresh Fission Products Following an Atomic Detonation," October 1953, LA-1597, 3.

18. Richard Greening Hewlett and Jack M. Holl, *Atoms for Peace and War, 1953–1961: Eisenhower and the Atomic Energy Commission* (Berkeley: University of California Press, 1989), 54.

19. US Congress, Joint Hearings, US Senate, Committee on the Judiciary, Senate Research Subcommittee of the Labor and Human Resources Committee, and House of Representatives, House Committee on Interstate and Foreign Commerce, Subcommittee on Oversight and Investigations, *Health Effects of Low-Level Radiation*, 96th Cong., 1st sess., April 19, 1979, vol. 1, 258 (statement of Dr. Harold A. Knapp).

20. *Allen v. United States*, 588 F. Supp. 247, 375–376.

21. Harold A. Knapp, "Observed Relations between the Deposition Level of Fresh Fission Products from Nevada Tests and the Resulting Levels of I-131 in Fresh Milk," March 1, 1963, files of US House of Representatives Commerce Committee on Oversight and Investigations.

22. Memorandum from Dr. Charles L. Dunham, Director, Division of Biology and Medicine, AEC, to Harold A. Knapp, October 24, 1962, files of US House of Representatives Commerce Committee on Oversight and Investigations.

23. Anna Almedrala, "Why You Should Stop Calling Thyroid Cancer 'Good' Cancer," *Healthy Living*, February 6, 2015 (quoting Alan Ho, MD, PhD).

6. HANFORD: SIGNS OF TROUBLE DOWNWIND

1. Karen Dorn Steele, "The Night the 'Little Demons' Were Born," *Spokesman-Review* (Spokane, WA), July 28, 1985, A10.

2. Michael D'Antonio, *Atomic Harvest: Hanford and the Lethal Toll of America's Nuclear Arsenal* (New York: Crown Publishers, 1991), 73–74.

3. Steele, "The Night the 'Little Demons' Were Born."

4. D'Antonio, *Atomic Harvest*, 73.

5. Steele, "The Night the 'Little Demons' Were Born."

6. "Rem," from "roentgen equivalent man," is defined as the dosage in rads that will cause the same amount of biological injury as one rad of x-rays or gamma rays: https://www.britannica.com/science/rem-unit-of-measurement.

7. United Nations Scientific Committee on the Effects of Atomic Radiation, "Malformations of the Eye," *Sources and Effects of Ionizing Radiation: 1977 Report to the General Assembly with Scientific Annexes*, https://www.unscear.org/docs/publications/1977/UNSCEAR_1977_Report.pdf.

8. John S. Reif, "Animal Sentinels for Environmental and Public Health," *Public Health Report* Suppl. 1 (2011): 126.

9. Iral C. Nelson and R. F. Foster, "Ringold: A Hanford Environmental Study, Internal Dosimetry Radiation Protection Operation," April 3, 1964, HW 78262 (REV), 3.

10. "Exposure Data of Ringold Study," January 5, 1963, PNL 10326, DEL, https://www.osti.gov/opennet/servlets/purl/16126430.pdf.

11. Nelson and Foster, "Ringold: A Hanford Environmental Study," 6.

12. W. C. Roesch, R. C. McCall, and H. E. Palmer, "Hanford Whole Body Counter 1959 Activities," December 1960, HW 67045, 38, https://www.osti.gov/opennet/detail?osti-id=16447907.

13. "Exposure Data of Ringold Study."

14. Roesch, McCall, and Palmer, "Hanford Whole Body Counter 1959 Activities," 4.

15. Nelson and Foster, "Ringold: A Hanford Environmental Study," 3.

16. Kate Brown, *Plutopia: Nuclear Families, Atomic Cities, and the Great Soviet and American Plutonium Disasters* (Oxford, UK: Oxford University Press, 2013), 247.

17. Matt Wood, "Understanding Radiation: Half Life," Tsukuba Science, April 3, 2011, http://www.tsukubascience.com/2011/04/understanding-radiation-half-life.

18. Brown, *Plutopia*, 248, citing "Letter to Subject" and memo to A. R. Keene, December 14, 1962, PPR, PNL-10335; "Status of Columbia River Environmental Studies for Hanford Works Area," July 31, 1961, DOE Germantown, RG 326, 1360, 3.

19. Brown, *Plutopia*, 248.

20. Nelson and Foster, "Ringold: A Hanford Environmental Study," 12.

21. Nelson and Foster, "Ringold: A Hanford Environmental Study," 10.

22. "Special Studies Conducted by Internal Dosimetry during 1962," March 19, 1963, 2, PNL 9709, DOE Public Reading Room, Richland, WA.

23. "Special Studies," 2.

24. J. K. Soldat and J. F. Honstead, "Dietary Levels for Tri-City Elementary School Children," February 26, 1968, BNWL-CC-1565, https://www.osti.gov/opennet /servlets/purl/16289244.pdf.

7. NTS: SIGNS OF TROUBLE DOWNWIND

1. Joint Committee on Atomic Energy, *Fallout from Nuclear Weapons Tests*, Summary of Hearings, 86th Cong., 1st sess., August 1959.

2. US Congress, US Senate Special Hearing, Subcommittee of the Committee on Appropriations, *Radioactive Fallout from Nuclear Testing at Nevada Test Site, 1950–1960*, 105th Cong., 1st sess., 1998 (hereafter cited as *Radioactive Fallout from Nuclear Testing*), 21 (testimony of Joseph L. Lyon, MD).

3. Edward S. Weiss, "Leukemia Mortality in Southwestern Utah, 1950–1964" (Washington, DC: US Public Health Service, 1965).

4. *Radioactive Fallout from Nuclear Testing*, 20 (testimony of Joseph L. Lyon).

5. Memorandum from Dwight A. Ink, Assistant General Manager, AEC, to AEC Chairman Seaborg, September 9, 1965, quoted in US Congress, House Committee on Interstate and Foreign Commerce, Subcommittee on Oversight and Investigations, *The Forgotten Guinea Pigs: A Report on Health Effects of Low-Level Radiation Sustained as a Result of the Nuclear Weapons Testing Program Conducted by the United States Government*, 96th Cong., 2nd sess., August 1980, 15 (hereafter cited as *Forgotten Guinea Pigs*).

6. Gordon Eliot White, "U.S. Kept Ignoring Evidence about Fallout's Deadly Effects," *Deseret News* (Salt Lake City, UT), October 28, 1990.

7. See US Congress, Joint Hearings, US Senate, Committee on the Judiciary, Senate Research Subcommittee of the Labor and Human Resources Committee, and House of Representatives, Committee on Interstate and Foreign Commerce, Subcommittee on Oversight and Investigations, *Health Effects of Low-Level Radiation*, 96th Cong., 1st sess., April 19, 1979, vol. 1 (hereafter cited as *Health Effects of Low-Level Radiation*).

8. *Radioactive Fallout from Nuclear Testing*, 21 (testimony of Joseph L. Lyon).

9. *Radioactive Fallout from Nuclear Testing*, 21 (testimony of Joseph L. Lyon).

10. Howard Ball, *Justice Downwind: America's Atomic Testing Program in the 1950s* (New York: Oxford University Press, 1986), 111.

11. Joseph L. Lyon, Melville R. Klauber, John W. Gardner, and King S. Udall, "Childhood Leukemias Associated with Fallout from Nuclear Testing," *New England Journal of Medicine* 300, no. 8 (March 1979): 397–402.

12. Ball, *Justice Downwind*, 113.

13. Lyon et al., "Childhood Leukemias."

14. *Radioactive Fallout from Nuclear Testing*, 21 (testimony of Joseph L. Lyon).

15. Gordon Eliot White, "Deaths High in Utah Fallout Area," *Deseret News*, August 12, 1977.

16. White, "Deaths High in Utah."

17. Gordon Eliot White, "Leukemia Reports Never Published," *Deseret News*, November 13, 1987.

18. White, "Leukemia Reports Never Published."

19. US Congress, House Committee on Interstate and Foreign Commerce, House Subcommittee on Health and the Environment, *Effects of Radiation on Human Health: Health Effects of Ionizing Radiation*, hearings, 95th Cong., 2nd sess., 1978 (hereafter cited as *Effects of Radiation on Human Health*); *Health Effects of Low-Level Radiation*.

20. *Effects of Radiation on Human Health*.

21. *Forgotten Guinea Pigs*.

22. *Forgotten Guinea Pigs*, 22.

23. Associated Press, "Children's Cancer Blamed on Fallout," *Oregonian* (Portland, OR), January 1, 1981, 22.

24. *Radioactive Fallout from Nuclear Testing*, 22 (testimony of Joseph L. Lyon).

25. *Radioactive Fallout from Nuclear Testing*, 22 (testimony of Joseph L. Lyon).

26. Walter Stevens, Duncan C. Thomas, Joseph L. Lyon, John E. Till, Richard A. Kerber, Steven L. Simon, Ray D. Lloyd, Naima Abd Elghany, and Susan Preston-Martin, "Leukemia in Utah and Radioactive Fallout from the Nevada Test Site: A Case-Control Study," *Journal of the American Medical Association* 264, no. 5 (September 1990): 585–591.

27. *Radioactive Fallout from Nuclear Testing*, 22 (testimony of Joseph L. Lyon).

28. C. J. Johnson, "Cancer Incidence in an Area of Radioactive Fallout Downwind from the Nevada Test Site," *Journal of the American Medical Association* 251, no. 2 (January 1984): 230–236.

29. S. L. Simon, R. D. Lloyd, J. E. Till, H. A. Hawthorne, D. C. Gren, M. L. Rallison, and W. Stevens, "Development of a Method to Estimate Thyroid Dose from Fallout Radioiodine in a Cohort Study," *Health Physics* 59, no. 5 (December 1990): 669–691.

30. Joseph L. Lyon, Stephen C. Alder, Mary Bishop Stone, Alan Scholl, James C. Reading, Richard Holubkov, Xiaoming Sheng, George L. White, Jr., Kurt T.

Hegmann, Lynn Anspaugh, F. Owen Hoffman, Steven L. Simon, Brian Thomas, Raymond Carroll, and A. Wayne Meikle, "Thyroid Disease Associated with Exposure to the Nevada Nuclear Weapons Test Site Radiation: A Reevaluation Based on Corrected Dosimetry and Examination Data," *Epidemiology* 17, no. 6 (November 2006): 604–614, 612.

31. Richard A. Kerber, John E. Till, Steven L. Simon, Joseph L. Lyon, Duncan C. Thomas, Susan Preston-Martin, Marvin L. Rallison, Ray D. Lloyd, and Walter Stevens, "A Cohort Study of Thyroid Disease in Relation to Fallout from Nuclear Weapons Testing," *Journal of the American Medical Association* 270, no. 17 (November 1993): 2076–2082.

32. *Radioactive Fallout from Nuclear Testing*, 22 (testimony of Joseph L. Lyon).

33. "Estimated Exposures and Thyroid Doses Received by the American People from Iodine-131 in Fallout Following Nevada Atmospheric Nuclear Bomb Tests," National Cancer Institute, 1997, https://www.cancer.gov/about-cancer /causes-prevention/risk/radiation/i131-report-.

34. I-131 Thyroid Dose and Risk Calculator for Nuclear Weapons Fallout for the US Population, National Cancer Institute, July 2003, https://radiationcalculators .cancer.gov/fallout.

35. Lyon et al., "Thyroid Disease Associated with Exposure."

36. *Health Effects of Low-Level Radiation*, vol. 1, 98.

37. US Congress, House of Representatives, Committee on the Judiciary, Subcommittee on Administrative Law and Governmental Relations, 101st Cong., 1st sess., *Discretionary Function Exemption of the Federal Tort Claims Act and the Radiation Exposure Compensation Act*, November 1 and 8, 1989, 326 (hereafter cited as *Discretionary Function Exemption*).

38. Leo K. Bustad, L. A. George, Jr., S. Marks, D. E. Warner, C. M. Barnes, K. E. Herde, H. A. Kornberg, and H. M. Parker, "Biological Effects of Iodine-131 Continuously Administered to Sheep," *Journal of Radiation Research*, vol. 6, no. 3 (March 6, 1957): 380–413, 397.

39. *Discretionary Function Exemption*, 327.

40. *Discretionary Function Exemption*, 320.

41. *Bulloch v. United States*, 474 US 1086 (1986), *denying cert. to*, 145 F. Supp. 824 (D. Utah 1956), *vacated*, 95 F.R.D. 123 (D. Utah 1982), *rev'd*, 721 F. 2d 713 (10th Cir. 1983), *rev'd on rehearing*, 763 F. 2d 1115 (10th Cir. 1985) [hereafter referred to as *Bulloch II*].

42. Philip K. Fradkin, *Fallout: An American Nuclear Tragedy* (1989; repr., Boulder, CO: Johnson Books, 2004), 234.

43. *Bulloch v. United States*, 721 F. 2d 713, 719 (10th Cir. 1983).

44. *Bulloch II*.

45. US Congress, House Committee on Interstate and Foreign Commerce, Subcommittee on Oversight and Investigations, *Low-Level Radiation Effects on Health*, 96th Cong., 1st sess., April 23, May 24, August 1, 1979 (hereafter cited as *Low-Level Radiation Effects on Health*), Appendix E, Annex 33, letter of January 29, 1980, from Harold A. Knapp to Dr. Leo K. Bustad, 975 et seq.

46. L. K. Bustad, S. Marks, N. L. Dockum, D. R. Kalkwarf, and H. A. Kornberg, *A Comparative Study of Hanford and Utah Range Sheep*, November 30, 1953, HW 30119, 43 (Conclusion no. 2).

47. Radiological Sciences Department, Quarterly Progress Report January–March 1952, April 16, 1952 HW 24131, 12.

48. Bustad et al., *A Comparative Study*, 43 (Conclusion no. 4).

49. Bustad et al., "Biological Effects of Iodine 131."

50. *Low-Level Radiation Effects on Health*, 1182 (Knapp regarding HW 30119 Conclusion no. 4). See also observations of Utah sheep ranchers regarding lambs born with grotesque deformities, *Discretionary Function Exemption*, 350.

51. Bustad et al., *A Comparative Study*, 43 (Conclusion no. 5).

52. *Low-Level Radiation Effects*, 1183 (Knapp re Bustad Conclusion no. 5).

53. *Low-Level Radiation Effects on Health*, 518 (Knapp, "Sheep Deaths in Utah and Nevada Following the 1953 Nuclear Tests, August 1, 1979").

54. *Low-Level Radiation Effects on Health*, 985, E-33-29 (Harold A. Knapp to Leo K. Bustad, letter dated 29 January 1980).

55. *Low-Level Radiation Effects on Health*, 985, E-33-29 (Knapp to Bustad).

56. *Low-Level Radiation Effects on Health*, 987, E-33-31 (Knapp to Bustad).

57. *Low-Level Radiation Effects on Health*, 987, E-33-31 (Knapp to Bustad).

58. *Low-Level Radiation Effects on Health*, 991, E-33-35 (Knapp to Bustad).

59. See William A. Fletcher, "Atomic Bomb Testing and the Warner Amendment: A Violation of the Separation of Powers," *Washington Law Review and State Bar Journal* 65 (January 1, 1990): 285–321, 286.

60. Under the FTCA, the United States has waived sovereign immunity, with twelve exceptions, including the "discretionary function exception," relating to claims based on "the exercise or performance or the failure to exercise or perform a discretionary function or duty on the part of the federal agency or an employee of the government, whether or not the discretion involved be abused," 28 USC § 2680(a).

61. Fletcher, "Atomic Bomb Testing," 304.

62. Fletcher, 306.

63. Fletcher, 307.

64. The Warner Amendment is also known as the Atomic Testing Liability Act, 50 USC § 2783.

65. See A. Constandina Titus and Michael W. Bowers, "Konizeski and the Warner Amendment: Back to Ground Zero for Atomic Litigants," *Brigham Young University Law Review* (1988): 387–408.

66. 50 USC § 2783, "Atomic Testing Liability Act" (discussion of procedure for removal of atomic testing cases to federal court under the Warner Amendment).

67. 28 USCA §§ 1346, 2671–2680 (West 1976 & Supp. 1989).

68. 28 USC § 2402.

69. *Irene Allen v. United States*, 588 F. Supp. 247 (D. Utah 1984) (hereafter cited as *Allen v. United States*).

70. 28 USC § 1346, 2671 et seq. (1988). Wrongful conduct or negligence is required under the Federal Tort Claims Act.

71. Thomas G. Alexander, "Radiation Death and Deception," historytogo .utah.gov.

72. "More probable than not" (by a preponderance of the evidence) refers to the burden of persuasion that the plaintiff must satisfy in a civil tort action.

73. Philip K. Fradkin, *Fallout: An American Nuclear Tragedy* (Tucson, AZ: University of Arizona Press, 1989, Boulder, CO: Johnson Books, 2004), 165.

74. *Allen v. United States*, 588 F. Supp. 247, 427 (hyperlink to Table 45).

75. See, e.g., "Tort," Legal Information Institute, https://www.law.cornell.edu /wex/tort.

76. *Allen v. United States*, 588 F. Supp. 247, 405–406.

77. *Allen v. United States*, 588 F. Supp. 247, 406.

78. *Allen v. United States*, 588 F. Supp. 247, 406.

79. See David Rosenberg, "The Causal Connection in Mass Exposure Cases: A 'Public Law' Vision of the Tort System," *Harvard Law Review* 97 (February 1984): 849–929, 851.

80. See, e.g., "But-for test," Legal Information Institute, https://www.law.cor nell.edu/wex/but-for_test.

81. Note (2015) on "Causation in Environmental Law: Lessons from Toxic Torts," *Harvard Law Review* 128 (March 20, 2019): 2256.

82. *In re Hanford Nuclear Reservation Litigation*, 521 F.3d 1028 (9th Cir. 2008).

83. *Allen v. United States*, 588 F. Supp. 247, 415.

84. *Allen v. United States*, 588 F. Supp. 247, 412.

85. Daniel M. Conway-Jones, "Factual Causation in Toxic Tort Litigation: A Philosophical View of Proof and Certainty in Uncertain Disciplines," *University of Richmond Law Review* 35, no. 4 (March 3, 2010): 875–941, 885.

86. See, e.g., *Summers v. Tice*, 33 Cal.2d 80 (1948).

87. *Allen v. United States*, 588 F. Supp. 247, 411.

88. *Allen v. United States*, 588 F. Supp. 247, 415.

89. *Allen v. United States*, 588 F. Supp. 247, 421.

90. *Allen v. United States*, 588 F. Supp. 247, 427 (Table 45). Comparison of bell-wethers' organ doses calculated by ORERP, the DOE dose assessment project (in rads); by Dr. John Gofman for the plaintiffs (in rads); by Beck and Krey for the government (in roentgens); and by Tamplin for the plaintiffs (in roentgens).

91. *Allen v. United States*, 588 F. Supp. 247, 385.

92. Fradkin, *Fallout*, 49.

93. *Allen v. United States*, 588 F. Supp. 247, 380.

94. *Allen v. United States*, 588 F. Supp. 247, 381.

95. *Allen v. United States*, 588 F. Supp. 247, 426. The Three Mile Island accident in 1979 involved the partial meltdown of the Unit 2 reactor near Middletown, PA.

96. *Allen v. United States*, 588 F. Supp. 247, 429.

97. In assessing the causal relationship between each bellwether's exposure to fallout and cancer or other disease, Judge Jenkins relied upon epidemiological studies, the United Nations Scientific Committee on the Effects of Atomic Radiation (UNSCEAR) report (1977), and the Biological Effects of Ionizing Radiation (BEIR) III report. See, e.g., *Allen v. United States*, 588 F. Supp. 247, 437–438.

98. *Allen v. United States*, 588 F. Supp. 247, 427–428 (Table 16, hyperlink to Table 45).

99. *Allen v. United States*, 588 F. Supp. 247, 428.

100. Ball, *Justice Downwind*, 162.

101. *Allen v. United States*, 588 F. Supp. 247, 427, n180.

102. *Allen v. United States*, 588 F. Supp. 247, 385.

103. *Discretionary Function Exemption*, 102 (statement of Sen. Orrin Hatch).

104. *Allen v. United States*, 588 F. Supp. 247, 381.

105. *Allen v. United States*, 588 F. Supp. 247, 260.

106. The term "discretionary function" was not defined within the FTCA. The intent of the FTCA was to waive sovereign immunity in order to allow citizens to sue the US government for torts committed by government employees within the scope of their employment. There are thirteen exceptions to this waiver of sovereign immunity, one of which is the discretionary function exception. The federal courts have been attempting to define this sovereign immunity–based defense since the passage of the FTCA in 1946. *Allen v. United States*, 816 F. 2d 1417 (10th Cir. 1987) at [20].

107. *Allen v. United States*, 816 F. 2d 1417.

108. *Allen v. United States*, 588 F. Supp. 247, 446 (hyperlink to Table 50).

109. *Allen v. United States*, 588 F. Supp. 247, 448. No guidance was included within the judgment for the more than one thousand remaining claims.

110. *Allen v. United States*, 816 F. 2d 1417, 1427 (J. McKay concurring).

111. *Allen v. United States*, 816 F. 2d 1417, 1427.

112. *Allen v. United States*, 108 S. Ct. 694 (1988).

113. A petition for a writ of certiorari (sometimes called a "cert petition") is a petition that a losing party may file with the US Supreme Court asking the court to review the decision of a lower appellate court.

114. Karen Dorn Steele, "Radiation Lawsuit Abandoned: Court Ruling Hurt Downwinders' Case," *Spokesman-Review* (Spokane, WA), February 5, 1988, 1.

115. Steele.

116. Steele.

8. HANFORD: THE SILENT HOLOCAUST

1. The N reactor, operating from 1963 to 1987, was the last of Hanford's nine plutonium production reactors and its only dual-purpose reactor, producing electricity as well as plutonium.

2. One of five chemical processing plants at Hanford, the PUREX plant chemically extracted plutonium from irradiated fuel rods. From 1956 to 1972 and from 1983 to 1988, the PUREX plant processed about 75 percent of the plutonium produced at Hanford.

3. "Radioactive Thorium Leak at Plutonium Plant No Cause for Alarm," *Seattle Times*, February 7, 1984.

4. "Silent Holocaust," sermon by Rev. Dr. William H. Houff, Unitarian Universalist Church of Spokane (excerpt reprinted with permission).

5. Michael D'Antonio, *Atomic Harvest: Hanford and the Lethal Toll of America's Nuclear Arsenal* (New York: Crown Publishers, 1991), 74.

6. D'Antonio, 75.

7. Tim Connor, "Nuclear Workers at Risk," *Bulletin of the Atomic Scientists* 46, no. 7 (1990): 24–28, 27.

8. Connor.

9. Keith Schneider, "U.S. Studies Health Problems near Weapon Plant," *New York Times*, October 17, 1988.

10. Michele Stenejhem Gerber, *On the Home Front: The Cold War Legacy of the Hanford Nuclear Site* (Lincoln: University of Nebraska Press, 2002), 3.

11. Karen Dorn Steele, "Hanford's Bitter Legacy," *Bulletin of the Atomic Scientists* 44, no. 1 (January–February 1988): 17–23, 20.

12. D'Antonio, *Atomic Harvest*, 126.

13. Steele, "Hanford's Bitter Legacy," 23.

14. Steele, 22.

15. Steele, 22.

16. The explosion in reactor no. 4 at Chernobyl took place on April 26, 1986, but was not detected until April 28 by scientists at the Forsmark nuclear plant a couple of hours north of Stockholm, Sweden. At first, the Soviets denied that anything had happened, but they finally admitted to the accident when Sweden warned that it was going to file an official alert with the International Atomic Energy Authority.

17. D'Antonio, *Atomic Harvest*, 131.

18. Steele, "Hanford's Bitter Legacy," 22. The N reactor was shut down in 1987 as a result of public concern following the Chernobyl accident.

19. See, e.g., US Congress, US Senate, Committee on Energy and Natural Resources, *The Chernobyl Accident and Implications for the Domestic Nuclear Industry*, 99th Cong., 2nd sess., June 19, 1986.

20. Indemnification agreements between Hanford contractors and the federal government under the Price-Anderson Nuclear Industries Indemnity Act (P-AA) would pay any attorneys' fees and any judgments obtained against contractors for injury downwind of Hanford, 42 USC §2210. See Chapter 9 for further discussion of the P-AA and the downwinders' personal injury litigation.

21. Karen Dorn Steele, "Radiation Study Set Up as Defense, Records Show," *Spokesman-Review* (Spokane, WA), February 23, 2005.

22. Steele.

23. "Hanford Environmental Dose Reconstruction Project (HEDR)—Richland, Washington," Radiation and Your Health, Centers for Disease Control and Prevention, https://www.cdc.gov/nceh/radiation/brochure/profile_hanford.htm.

24. See Scott Davis, Kenneth J. Kopecky, Thomas E. Hamilton, Lynn Onstad, and the Hanford Thyroid Disease Study Team, "Thyroid Neoplasia, Autoimmune Thyroiditis, and Hypothyroidism in Persons Exposed to Iodine 131 from the Hanford Nuclear Site," *Journal of the American Medical Association* 292, no. 21 (December 1, 2004): 2600–2613.

25. Davis et al.

26. "Biologically significant radionuclides" refers to radioactive substances, such as plutonium or radioactive iodine, that provide the most significant health hazards to humans among all of the radioactive substances released from human activity.

27. C. M. Heeb, S. P. Gydesen, J. C. Simpson, and D. J. Bates, "Reconstruction of Radionuclide Releases from the Hanford Site, 1944–1972," *Health Physics* 71, no. 4 (October 1996): 545–555.

28. Keith Schneider, "Radiation Peril at Hanford Is Detailed," *New York Times*, July 13, 1990, A8 (statement of Allen W. Conklin, radiation specialist with the Washington State Department of Health).

29. Hanford Thyroid Disease Study Final Report, June 21, 2002, Executive Summary, xxxiv, https://www.cdc.gov/nceh/radiation/hanford/htdsweb/pdf/htds report.pdf.

30. National Academy of Sciences, *Review of the Hanford Thyroid Disease Study Draft Final Report*, National Academy of Sciences, Committee on an Assessment of CDC Radiation Studies from DOE Contractor Sites: Subcommittee to Review the HTDS Final Results and Report, Board on Radiation Effects Research, Commission of Life Sciences, Washington, DC, 2000 (hereafter cited as NAS, *Review of the Hanford Thyroid Disease Study*). Had HTDS found a dose response, it would have meant that the study participants with higher HEDR doses had higher incidences of thyroid disease and thyroid cancer than those within the cohort with lower HEDR doses.

31. Matthew L. Wald, "No Radiation Effect Found at Northwest Nuclear Site," *New York Times*, January 28, 1999.

32. Scott Davis and Kenneth Kopecky, "Executive Summary: HTDS Draft Final Report," Fred Hutchinson Cancer Research Center, Seattle, Washington, 18.

33. Letter from citizen groups to Richard Jackson, MD, CDC, February 18, 1999, objecting to the manner in which HTDS results were presented and requesting an extended peer review of the study.

34. Karen Dorn Steele, "Fallout from Thyroid Study: Critics Fault CDC for Early Release of Hanford Results, Unreviewed Research," *Spokesman-Review*, February 14, 1999.

35. A retrospective cohort study is a study of a group (cohort) of people who already have developed the diseases of interest.

36. CDC, *Summary of the Preliminary Results—The Hanford Thyroid Disease Study Draft Final Report*, May 1999, 4.

37. Tom Reynolds, "Final Report of Hanford Thyroid Disease Study," *Journal of the National Cancer Institute* 94, no. 14 (July 17, 2002): 1046–1048.

38. Fred Hutchinson Cancer Research Center, "Preliminary Technical Review of the Hanford Thyroid Disease Study Draft Final Report," 1998, 6 (hereafter cited as "Preliminary Review of HTDS").

39. "Preliminary Review of HTDS," 6.

40. Brett Sholtis, "Thyroid Cancer Study Re-Ignites Debate over Three Mile Island Accident," Allegheny Front, March 22, 2019, https://www.alleghenyfront.org/thyroid-cancer-study-re-ignites-debate-over-three-mile-island-accident.

41. Solveig Torvik, "Study Further Muddies Hanford Waters," *Seattle Times*, February 28, 1999.

42. See NAS, *Review of the Hanford Thyroid Disease Study*.

43. The National Research Council of the NAS conducted the review. Its work is funded by the CDC.

44. *Hanford Health Information Network Final Report* (June 2000), *Oregon HHIN Program Final Report*, courtesy of Washington State Digital Archives, Cheney, WA.

45. Testimony of Sherry Katherine Dunn, RN, BSN, Oregon state health educator with the HHIN, at a public hearing before the National Research Council, Board on Radiation Effects Research, Review of the Hanford Thyroid Disease Study, Spokane, WA, June 19, 1999.

46. Joseph L. Lyon, Stephen C. Alder, Mary Bishop Stone, Alan Scholl, James C. Reading, Richard Holubkov, Xiaoming Sheng, George L. White, Jr., Kurt T. Hegmann, Lynn Anspaugh, F. Owen Hoffman, Steven L. Simon, Brian Thomas, Raymond Carroll, and A. Wayne Meikle, "Thyroid Disease Associated with Exposure to the Nevada Nuclear Weapons Test Site Radiation: A Reevaluation Based on Corrected Dosimetry and Examination Data," *Epidemiology* 17, no. 6 (November 2006): 604–614.

47. NAS, *Review of the Hanford Thyroid Disease Study*, 31.

48. NAS, 30–31.

49. NAS, 31.

50. See James A. Ruttenber et al., "A Technical Review of the Final Report of the Hanford Thyroid Disease Study" (expert report prepared for the Hanford litigation), March 30, 2004, 3.

51. Kate Brown, *Manual for Survival: A Chernobyl Guide to the Future* (New York: W. W. Norton, 2019), 69.

52. See Chapter 9 for discussion of bellwether jury trials in *In re Hanford Nuclear Reservation Litigation*.

53. Panel discussion: "Hanford Releases: Hanford Thyroid Disease Study, Individual Dose Assessment, and Next Steps," November 2, 1999 (statement of Kathleen S. Fox-Williams, Washington State Department of Health).

54. Lyon et al., "Thyroid Disease Associated with Exposure."

55. Lyon et al., 613.

56. Lyon et al., 613.

57. W. Stevens, J. E. Till and D. C. Thomas, *Assessment of Leukemia and Thyroid Disease in Relation to Fallout in Utah: Report of a Cohort Study of Thyroid Disease and Radioactive Fallout from the Nevada Test Site* (Salt Lake City: University of Utah, 1992).

58. Lyon et al., "Thyroid Disease Associated with Exposure," 613.

59. Lyon et al., 613.

60. Rudi L. Nussbaum, Patricia P. Hoover, Charles M. Grossman, and Fred D. Nussbaum, "Community-Based Participatory Health Survey of Hanford, WA,

Downwinders: A Model for Citizen Empowerment," *Society and Natural Resources* 17 (2004): 547–559, 551, 549.

61. Nussbaum et al., 554.

62. Nussbaum et al., 554.

63. Nussbaum et al., 554.

64. The Hanford Downwinders' Coalition was an advocacy group based in Seattle led by downwinder Judith Jurji.

65. JSI Center for Environmental Health Studies, *Report of R-11 Survey Results*, November 14, 1995. The comparison survey was the National Health Interview Survey conducted in 1993.

66. L. M. Tatham, Frank J. Bove, Wendy E. Kaye, and Robert F. Spengler, "Population Exposures to I-131 Releases from Hanford Nuclear Reservation and Preterm Birth, Infant Mortality, and Fetal Deaths," *International Journal of Hygiene and Environmental Health* 205, no. 1–2 (March 2002): 41–48.

67. See Nicole Blonio-Zorkin, Mariana Golts, and Virginia C. Fernandes, "Severe Hypothyroidism Presenting with Acute Mania and Psychosis: A Case Report and Literature Review," *Bipolar Disorder* 3 (2017): 116–119, https://www.longdom .org/open-access/severe-hypothyroidism-presenting-with-acute-mania-and -psychosis-a-casereport-and-literature-review-2472-1077-1000116.pdf.

68. Mark Ryan, "Brain Fog Matters," Hashimoto's Healing, https://www.hashi motoshealing.com/brain-fog-matters.

69. Ryan.

70. D. Geracioti, Jr., "Identifying Hyperthyroidism's Psychiatric Presentations," *Current Psychiatry* 5, no. 12 (December 2006): 84–92.

71. Ryan, "Brain Fog Matters."

72. Frank Lanzisera, "Thyroid Disease and Digestive Problems," *Total Health* magazine, http://www.totalhealthmagazine.com/Thyroid-Health/Thyroid-Disease -Digestive-Problems.html.

73. David Brownstein, "Hypothyroidism Causes Digestion Problems," Newsmax, http://www.newsmax.com/Health/Dr-Brownstein/hypythyroidismIBS-hydro chloric-acid-digestion/2014/06/04/id/575174.

74. Lanzisera, "Thyroid Disease and Digestive Problems."

75. C. Maser, Arnbjorn Toset, and Sanziana Roman, "Gastrointestinal Manifestations of Endocrine Disease," *World Journal of Gastroenterology* 12, no. 20 (May 28, 2006): 3174–3179, https://www.ncbi.nlm.nih.gov/pmc/articles/PMC4087959.

76. See R. Daher, Thierry Yazbeck, Joe Bou Jaoude, and Bassam Abboud, "Consequences of Dysthyroidism on the Digestive Tract and Viscera," *World Journal of Gastroenterology* 15, no. 23 (June 21, 2009): 2834–2838, https://www.ncbi.nlm.nih .gov/pmc/articles/PMC2699000.

77. "Understanding Crohn's Disease," Crohn's & Colitis, https://www.crohns andcolitis.com/crohns?cid=ppc_ppd_ggl_cd_da_what_is_crohn%27s_disease _Exact_64Z1867745.

9. HANFORD DOWNWINDERS TURN TO THE COURTS

1. "Initial Hanford Radiation Dose Estimates," report of the Technical Steering Panel of the Hanford Environmental Dose Reconstruction Project, 1990.

2. A "public liability action," as defined within the Atomic Energy Act (AEA), 42 USC § 2210, is any suit asserting public liability growing out of exposure to nuclear radiation (see 42 USC § 2014(hh)). "Public liability" is any legal liability arising out of or resulting from a nuclear incident or evacuation (42 USC § 2014(w)).

3. The Price-Anderson Act, § 170 of the Atomic Energy Act (42 USC § 2210).

4. American Nuclear Society, "The Price-Anderson Act: Background Information," http://www.ans.org/pi/ps/docs/ps54-bi.pdf.

5. AEA, 42 USC SS 2014(q).

6. In re Hanford Nuclear Reservation Litigation, 780 F. Supp. 1551, 1555 (E.D. Wash 1991).

7. In re Hanford Nuclear Reservation Litigation, 1998 WL 775340, *2.

8. Plaintiffs' Consolidated Complaint, In re Hanford Nuclear Reservation Litigation, 780 F. Supp. 1551 (E.D. Wash 1991), 60–76, courtesy of University of Washington Libraries, Special Collections, Hanford Litigation Records.

9. In re Hanford Nuclear Reservation Litigation, 1998 WL 775340, *2.

10. The P-AA distinguishes between contractors operating commercial nuclear power plants for the US Nuclear Regulatory Commission (NRC) and contractors with the DOE. DOE contractors, including Hanford contractors, are generally involved in the testing or production of nuclear weapons. Both types of contractors receive the same limit on liability under the P-AA, but only DOE contractors are indemnified by the government (AEA, 42 USC § 2210 (c), § 2210 (d)).

11. For a discussion of generic and specific causation in toxic tort litigation, see David E. Bernstein, "Getting to Causation in Toxic Tort Cases," Brooklyn Law Review 74, no. 1 (Fall 2008): 51–74.

12. In re Hanford Nuclear Reservation Litigation, 1998 WL 775340, *13.

13. See, e.g., "Overview: Doubling Dose," Oxford Reference, http://www .oxfordreference.com/view/10.1093/oi/authority.20110803095728405.

14. "Dismissed with prejudice" is a final judgment, and the plaintiff is barred from bringing an action on the same claim.

15. In re Hanford Nuclear Reservation Litigation, 1998 WL 775340, *70.

16. In re Hanford Nuclear Reservation Litigation, 292 F. 3d 1124, 1134 (9th Cir. 2002).

17. *In re Hanford Nuclear Reservation Litigation*, 292 F. 3d 1124, 1137.

18. *In re Hanford Nuclear Reservation Litigation*, 292 F. 3d 1124, 1136.

19. *In re Hanford Nuclear Reservation Litigation*, Order of Reference, CR 388, 7, courtesy of University of Washington Libraries, Special Collections, Hanford Litigation Records.

20. Associated Press, "Report: Scientist Disputes Hanford Radiation Studies," *Lewiston Tribune*, January 21, 1995.

21. Nicholas K. Geranios, "US Judge Pulls Out of Lawsuit on Hanford Radiation," *Seattle Times*, March 12, 2003.

22. Associated Press, "Radiation Report May Have Been Flawed: Hanford Residents Might Have Gotten Bigger Doses Than Study Showed," *Seattle Post-Intelligencer*, June 6, 2003.

23. Thomas H. Pigford, *Assessment of Radiation Dose Estimates Made by the Hanford Environmental Dose Reconstruction Project*, December 1994, courtesy of University of Washington Libraries, Special Collections, Hanford Litigation Records.

24. Jenna Greene, "In Hanford Saga, No Resolution in Sight," *National Law Journal*, June 20, 2011.

25. The HTDS found a surprising amount of thyroid disease downwind of Hanford, although its prevalence was not dose related. See Trisha T. Pritikin, "Insignificant and Invisible: The Human Toll of the Hanford Thyroid Disease Study," in *Tortured Science: Health Studies, Ethics and Nuclear Weapons in the United States*, ed. Dianne Quigley, Amy Lowman, and Steve Wing (Amityville, NY: Baywood Publishing, 2012), 25–52, 36–37.

26. *In re Hanford Nuclear Reservation Litigation*, Order re: Plaintiffs' Motion for Summary Judgment: Abnormally Dangerous Activity, courtesy of University of Washington Libraries, Special Collections, Hanford Litigation Records.

27. *In re Hanford Nuclear Reservation Litigation*, 350 F. Supp. 2d 871, 888 (E.D. Wash 2004).

28. *In re Hanford Nuclear Reservation Litigation*, 350 F. Supp. 871, 875.

29. *In re Hanford Nuclear Reservation Litigation*, 350 F. Supp. 871, 879.

30. *Langan v. Valicopters, Inc.*, 88 Wash. 2d 855, 861, 567 P. 2d 218 (1977), quoting Restatement (Second) of Torts, § 520, comment f.

31. *In re Hanford Nuclear Reservation Litigation*, 350 F. Supp. 2d 871, 888.

32. *In re Hanford Nuclear Reservation Litigation*, 350 F. Supp. 2d 871, 888.

33. In *Allen*, strict liability had not been asserted by the plaintiffs, as strict liability cannot be the basis for a tort claim filed against the United States (*Dalehite v. United States*, 346 U.S. 14, 44–45 (1953)).

34. Warren Cornwall, "Hanford Likely Caused Cancer Downwind, Jury Decides," *Seattle Times*, May 20, 2005.

35. Associated Press, "Jury Awards $500M [sic] to Two Plaintiffs in Hanford Lawsuits," May 20, 2005, https://tdn.com/business/local/jury-awards-m-to-two-plaintiffs-in-hanford-lawsuits/article_802aae87-f014-5a5a-8088-d59271cd4440.html.

36. Ellen Sussman, "Hanford Puts Woman in Fight for Her Life," *Green Valley News* (Green Valley, AZ), February 28, 2009.

37. Nicholas K. Geranios, "Second Hanford Downwinder Trial Begins," *Seattle Times*, November 7, 2005.

38. "Downwinder Expected to Die within 2 Years," *Spokesman-Review* (Spokane, WA), November 8, 2005.

39. Annette Cary, "Decades-Long Hanford Downwinder Lawsuit Settles," *Tri-City Herald* (Kennewick, WA), October 7, 2015.

40. Karen Dorn Steele and Thomas Clouse, "Jury Rejects Rhodes' Lawsuit," *Spokesman-Review*, November 24, 2005.

41. Karen Dorn Steele, "Radiation Study Set Up as Defense, Records Show," *Spokesman-Review*, February 23, 2005.

42. Steele.

43. Steele and Clouse, "Jury Rejects Rhodes' Lawsuit."

44. Associated Press, "No New Trial for Hanford Downwinder," *Seattle Post-Intelligencer*, January 22, 2006.

45. Chelsea Bannac, "Deputies Say Deaths May Be Suicides," *Spokesman-Review*, December 15, 2010.

46. Cary, "Decades-Long Hanford Downwinder Lawsuit Settles."

47. Cary.

48. Greene, "In Hanford Saga, No Resolution in Sight."

49. Greene.

50. Cary, "Decades-Long Hanford Downwinder Lawsuit Settles."

51. In re Hanford Nuclear Reservation Litigation, Arbitrator's Memorandum Report Re: Arbitrator Awards to Hanford Nodule Surgery/Nonsurgery Claimants, November 17, 2015.

52. Annette Cary, "Final Hanford Downwinder Lawsuit Settled after 24 Years," *Tri-City Herald*, October 8, 2015.

53. Hanford Health Information Network, "Hot Spots: Weather and Hanford's Radiation Releases to the Air," Washington State Digital Archives, Cheney, WA.

54. Alan B. Benson, *Hanford Radioactive Fallout: Hanford's Radioactive Iodine-131 Releases (1944–1956)* (Cheney, WA: High Impact Press, 1989), 39.

55. J. K. Soldat, "Management of Radioactive Effluent Gases at Hanford Atomic Products Operation," August 7, 1956, 27, University of Washington Libraries, Special Collections, Hanford Litigation Records.

56. Benson, *Hanford Radioactive Fallout*, 39.

57. Hanford Health Information Network, "Hot Spots."

58. Kadlec was removed from AEC control in 1956.

10. REVERSAL OF *ALLEN*: THE CATALYST FOR CHANGE

1. "About the Federal Black Lung Program," US Department of Labor, Division of Coal Mine Workers' Compensation, https://www.dol.gov/owcp/dcmwc /mission.htm.

2. Howard Ball, *Justice Downwind: America's Atomic Testing Program in the 1950s* (New York, Oxford University Press, 1986), 178.

3. Ball, 181.

4. Ball, 192.

5. Holly Barker, M. Johnston, and Barbara Rose, "Seeking Compensation for Radiation Survivors in the Marshall Islands: The Contribution of Anthropology," *Cultural Survival Quarterly* 24, no. 1 (March 2000): 48–50. The Marshall Islands Tribunal typically paid out only 10–15 percent of the claims awarded before the money was gone.

6. US Congress, House of Representatives, Committee on the Judiciary, Subcommittee on Administrative Law and Governmental Relations, 101st Cong., 1st sess., *Discretionary Function Exemption of the Federal Tort Claims Act and the Radiation Exposure Compensation Act*, November 1 and 8, 1989, 96 (hereafter cited as *Discretionary Function Exemption*).

7. Radiation Exposed Veterans Compensation Act of 1988, 38 USC 101 note, https://www.govinfo.gov/content/pkg/STATUTE-102/pdf/STATUTE-102-Pg485 .pdf.

8. *Discretionary Function Exemption*, 97 (statement of Sen. Orrin Hatch).

9. Ball, *Justice Downwind*, 171.

10. *Allen v. United States*, 816 F.2d 1417, 1427 (10th Cir. 1987).

11. 42 USC § 2210 note.

12. J. A. Reissland, "BEIR III, 'The Effects on Populations of Exposure to Low Levels of Ionizing Radiation,'" *Journal of the Society for Radiological Protection* 1, no. 2 (Summer 1981): 17–22.

13. The place where a cancer starts is the "primary" site. A cancer that starts there is the "primary cancer" as opposed to cancer that has spread to other parts of the body, where it is a metastatic or secondary cancer.

14. "Clinical Guidelines," Health Resources and Services Administration, https://www.hrsa.gov/get-health-care/conditions/radiation-exposure/clinical .html.

15. Downwinder Claim Form, Radiation Exposure Compensation Program, US Department of Justice, Civil Division, https://www.justice.gov/sites/default/files /civil/legacy/2012/02/09/RECA_Downwinder.pdf.

16. Scott Szymendera, "The Radiation Exposure Compensation Act (RECA): Compensation Related to Exposure to Radiation from Atomic Weapons Testing and Uranium Mining," Library of Congress, Congressional Research Service, June 11, 2019. For a discussion of the problems facing downwinders seeking compensation under RECA, see, e.g., Marie I. Boutté, "Compensating for Health: The Acts and Outcomes of Atomic Testing," *Human Organization* 61, no. 1 (2002): 41–50.

17. Karen Dorn Steele, "Study Ties Fallout to US Increase in Thyroid Cancer: Five-Year Delay in Releasing Results Called 'Major Health Scandal,'" *Spokesman-Review* (Spokane, WA), July 24, 1997. Congress had requested the study fourteen years earlier, and fallout data collection for the nearly 3,100 counties included had been completed in 1992.

18. Matthew L. Wald, "Book Examines Nevada Test That Left Fallout in Troy, N.Y.," *New York Times*, April 18, 2003.

19. Karen Dorn Steele, "Children Caught Worst of Fallout: Inland West Counties Took Brunt of Radiation in 50s," *Spokesman-Review*, August 2, 1997.

20. The only additions to downwinder eligibility areas occurred through amendments to RECA in 2000. Amendments to RECA in 2002, passed as part of another bill, among other things, fixed the accidental deletion of certain geographic eligibility areas from the original act.

21. The latest proposed expansion, the Radiation Exposure Compensation Act Amendments of 2019, introduced on March 28, 2019, by a bipartisan coalition of western US senators, would make NTS downwinders in Idaho, Arizona, Colorado, Montana, Nevada, New Mexico, and Utah, as well as post-1971 uranium workers in New Mexico, eligible under RECA.

22. "Radiation Exposure Screening," RESEP, North County Healthcare, https:// northcountryhealthcare.org/community-programs/radiation-exposure -screening.

23. "Find a Clinic," Health Resources and Services Administration, https:// www.hrsa.gov/get-health-care/conditions/radiation-exposure/clinic.html.

24. The ATSDR was created in 1980 as part of Superfund legislation. Under Superfund, the ATSDR is required to initiate a health surveillance program if it determines that "there is significant increased risk of adverse health effects in humans from exposure to hazardous substances based on the results of a health assessment," 42 USC. § 9604(i)(9).

25. Karen Dorn Steele, "DOE Pulls Out of Press Briefing," *Spokesman-Review*, July 3, 1998.

26. Greg Thomas, ATSDR, "Hanford Releases, Hanford Thyroid Disease Study, Individual Dose Assessment, and Next Steps," November 2, 1999, 21, Washington State Digital Archives, Cheney, WA.

27. Steele, "DOE Pulls Out of Press Briefing."

28. "Hanford: Woman Appeals Dismissal of Radiation Lawsuit," *Kitsap Sun* (Bremerton, WA), June 6, 1999.

29. Northwest Environmental Education Foundation press notice, "Hanford Medical Monitoring Impasse: At Risk from Hanford's Exposures, Thousands Find the Clinic Door Is Locked," July 2, 1998.

30. Greg Thomas, ATSDR, "Panel Two: Hanford Releases, Hanford Thyroid Disease Study, Individual Dose Assessment, and Next Steps," November 2, 1999, 17–19, Washington State Digital Archives, Cheney, WA.

31. Comprehensive Environmental Response, Compensation and Liability Act of 1980, 42 USC §§ 9601–9675 (2000), commonly referred to as "Superfund."

32. See Superfund, citizen suits, 42 USC §§ 9659.

33. *Trisha T. Pritikin v. United States Department of Energy, et al.*, 47 F. Supp. 2d 1225 (E.D. Wash. 1999), affirmed at 254 F.3d 791 (9th Cir. 2001), cert. denied February 19, 2002, 534 US 1133; 122 S. Ct. 1076; 151 L.Ed.2d 977.

34. "Special Exposure Cohort Petition Information," US National Institute for Occupational Safety and Health (NIOSH) Radiation Dose Reconstruction Project, Hanford, CDC, https://www.cdc.gov/niosh/ocas/hanford.html#sec.

35. "Special Exposure Cohort," Cold War Patriots, https://coldwarpatriots.org/benefits/applying-for-benefits/special-exposure-cohort.

36. Dose reconstruction is carried out by NIOSH.

37. "FAQs: Probability of Causation," NIOSH, https://www.cdc.gov/niosh/ocas/faqspoc.html.

38. *In re Hanford Nuclear Reservation Litigation*, 521 F. 3d 1028 (9th Cir. 2008).

39. Marie I. Boutté, "Compensating for Health: The Acts and Outcomes of Atomic Testing," *Human Organization* 61, no. 1 (2002): 41–50, 42, citing Roy L. Brooks, ed., *When Sorry Isn't Enough: Controversy over Apologies and Reparations for Human Injustices* (New York: New York University Press, 1999).

40. See specified cancers under the EEOICPA, §§yy(5)(b), https://www.dol.gov/owcp/energy/regs/compliance/policyandprocedures/proceduremanual html/unifiedpm/unifiedpm_part0/chapter0–0500definitions.htm.

41. Szymendera, "The Radiation Exposure Compensation Act," 6.

42. H. M. Parker, "Status of Ground Contamination Problem," September 15, 1954, 6, HW 33068.

43. See W. A. McAdams, "A History of the Redox Ruthenium Problem," July 16, 1954 (HW 32473), University of Washington Libraries, Special Collections, Hanford Litigation Records.

44. Parker, "Status of Ground Contamination Problem."

45. Dennis J. Carroll, "Downwinders Welcome Study of Trinity Blast's Impacts," *New Mexican* (Santa Fe, NM), January 25, 2014.

ACKNOWLEDGMENTS

This project would not have been possible without the support of Tom Foulds of the Hanford Litigation Office in Seattle and Richard Eymann of Eymann Allison Hunter Jones in Spokane, two of the attorneys representing the downwinders in In re Hanford Nuclear Reservation Litigation. The two lawyers provided me with invaluable assistance through contacting former Hanford downwinder clients they thought might be willing to participate in this project. Additional thanks to my friend Jennifer Hunt, office manager and paralegal at the now shuttered Hanford Litigation Office, and to Diane Latta, former paralegal at Eymann Allison Hunter Jones, for their invaluable help and guidance throughout the process of client contact. I am additionally grateful that the words of Tom Foulds and Richard Eymann serve as the Foreword for this book.

I am honored that Karen Dorn Steele agreed to write the Introduction. Steele, the investigative journalist who courageously broke the Hanford story, is a true hero(ine) to the downwinders. Without Steele's steadfast determination to unearth Hanford's secrets and to chronicle the cancer and other disease reported in communities surrounding the facility, I am doubtful that Hanford's legacy of harm to downwind and downriver communities would have been made public by the DOE.

I feel privileged to have spent time with the individuals whose stories appear in this book. Many of these stories are heartbreaking. It is my hope that the power of our combined words, bringing to light through our personal stories the extent of harm caused to downwind and downriver communities by Hanford operations, will provide solace to all of those exposed to Hanford's secret off-site radiation releases and who now struggle with the burdens of radiogenic cancer and other serious radiogenic disease.

Over thirty years have passed since Hanford's irradiation of the people of the Pacific Northwest first became public knowledge in 1986. With the passage of time, the collective voice of the downwinders has been considerably weakened by the death of a number of outspoken advocates, including Jay Mullen and Marlene Campbell, whose stories are found within these pages. I miss Jay's wry sense of humor and his unrelenting advocacy over so many years on behalf of civilians injured by Hanford operations.

Marlene, another articulate voice for the downwinders, supplied me with frequent encouragement over several years and provided wonderful anecdotes about our shared experience as children downwind. I miss her.

Greatest thanks to Roger Briggs, MPH, DOE, Richland Operations Office, Quality, Safety and Health Division (1992–1996), Office of Special Concerns (1996–2000); Kate Brown, PhD, professor of science, technology and society, Massachusetts Institute of Technology; Jim Feldman, PhD, of the University of Wisconsin–Oshkosh; Robert (Bo) Jacobs, PhD, professor, Hiroshima Peace Institute and Graduate School of Peace Studies, Hiroshima City University, and codirector, Global Hibakusha Project; Barbara Rose Johnston, PhD, of the Center for Political Ecology; Michael Mays, PhD, director of the Hanford History Project and professor of history, Washington State University Tri-Cities; and Kenneth W. Pritikin, attorney-at-law, for review of all or part of several iterations of this manuscript.

The first-rate topographic map of the exposure residences of the Hanford downwinder plaintiffs whose stories are featured in this book was created by Matt Stevenson at Core GIS in Seattle. Thank you, Matt.

To Irene Lusztig, professor, film and digital media, University of California, Santa Cruz, and to Regina Longo, media archivist and lecturer, Brown University, my gratitude for the expert sleuthing that helped track down hard-to-find historical images. And to Robert Franklin, archivist at the Hanford History Project, thank you for your help securing historical photos from your collection. In my opinion, the talented archivists who help guide scholarly research within archival and special collections are no less than national treasures. I am very appreciative of the guidance and support of Anne Jenner, director of special collections, and her staff at my alma mater, the University of Washington, as I worked my way through the Hanford litigation records, an irreplaceable collection documenting, through plaintiff depositions, court filings, epidemiological studies, expert reports, and other records, the extent of human harm more likely than not caused by Hanford operations. These records were generously gifted to UW Special Collections in late 2015 by attorney Tom Foulds upon final settlement of the litigation. My appreciation as well for solid research assistance provided by Debbie Bahn, archivist; Whitney Waengaert, archives intern; and the staff of the Washington State Digital Archives, Cheney, Washington, where a substantial collection of Hanford downwinder papers and health records is housed.

I feel extremely fortunate that this project found a home with University Press of Kansas (UPK). Michael Briggs, thanks for the kind words and encouragement when an early version of the project first came across your desk a number of years ago. Kim Hogeland, former UPK acquisitions editor, thank you for believing in this project, for your kindness, and for excellent editorial guidance throughout. Also, greatest appreciation to UPK's David Congdon, Kelly Chrisman Jacques, Michael Kehoe, Andrea Laws, Larisa Martin, Erica Nicholson, and Connie Oehring for a production process that flowed flawlessly.

To Richard Pierson, attorney-at-law (ret.) and plaintiffs' counsel for In re Hanford, your support and kindness along the way helped greatly.

To my friend and agent extraordinaire Jody Rein, I am so thankful for your guidance, patience, and editorial expertise.

And to my family—husband, Ken; daughter, Siera; and son, Brendan—your love and support keep me strong.